Learning from VLSI Design Experience

Weng Fook Lee

Learning from VLSI Design Experience

Springer

Weng Fook Lee
Emerald Systems
Bayan Lepas, Malaysia

ISBN 978-3-030-03237-1 ISBN 978-3-030-03238-8 (eBook)
https://doi.org/10.1007/978-3-030-03238-8

Library of Congress Control Number: 2018960261

This Springer imprint is published by the registered company Springer Nature Switzerland AG
The registered company address is: Gewerbestrasse 11, 6330 Cham, Switzerland

Dedicated to my kids, my most treasured gift.

Preface

Digital systems are widely used in our daily lives. From household appliances to mobile smartphone and tablets, digital IC (integrated circuits) chips form a large part of our lives, whether we can clearly identify the presence of these IC chips in the systems we use. As technology becomes a bigger part of our lives, these IC chips increase in complexity from a few thousand transistors (Intel's 4004 in 1971) to 3.3 billion transistors (Apple's A10 Fusion in 2016). More and more functionalities are cramped into a single IC chip increasing its complexity. It is common to find IC chips with few million transistors and more. Such complexity increases the difficulty level for IC designers to design the circuitry for the IC chips. Digital design methodology and flow are used with Hardware Description Language such as Verilog to aid the designer to design these complex circuitries.

Chapter 1 is an introduction to digital systems, while Chap. 2 discusses about analog/custom design flow and digital design flow. Technical details on each step of the flow are discussed from RTL (Register-Transfer Level) coding through simulation, synthesis, auto place and route to final tapeout.

Chapter 2 covers details on the cells in a standard cell library. Standard cell library plays an important role in synthesis as the synthesis process synthesizes RTL code into logic gates using the available logic gates from the standard cell library. Logic cells commonly found in standard cell library are discussed with schematics on the standard cell's structure. Design constraints such as input delay, output delay, multicycle path, and false path are discussed in Chap. 2.

Clock tree synthesis which plays an important role in physical backend auto place and route is described in this chapter, describing how clock skew occurs between different clock paths. Floorplan of physical backend prior to clock tree synthesis is also discussed in Chap. 2.

Chapter 3 describes in detail on clock domain crossing for multi-asynchronous clocks in a design. Calculation of mean time and explanations using synchronizers in extending the mean time between failures for clock domain crossing are described. The conditions of using synchronizer to fix clock domain crossing for faster transmit clock compared to receiving clock and vice versa are discussed in detail in Chap. 3.

Differences between synchronous reset and asynchronous reset are also discussed in Chap. 3 with examples of correct RTL coding for these reset conditions.

Latch inference is undesirable in logic design and should be avoided. Chapter 4 discusses RTL coding conditions that lead to latch inference in a design. Examples of if-else statements and case statements that lead to latch inference are described in this chapter.

With majority of ASIC and SOC design being multimillion gates, testability of design becomes a critical issue. Chapter 5 discusses about scan chain and ATPG and how it can be used for testability. There are examples in Chap. 5 showing how a scan insertion occurs in a design with the scan flops being stitched to form scan chain. Test compression is also discussed in Chap. 5 on handling multiple scan chains for large design.

For large design with multiple asynchronous clocks, Chap. 5 discusses on handling of scan chain design when crossing clock domain. Capture cycle of ATPG involving multiple clocks is also covered in Chap. 5. Apart from scan chain and ATPG, Chap. 5 also covers logic BIST and memory BIST and how it is used in ASIC and SOC.

Signed Verilog is important for design that requires computation for positive and negative, for example, field-oriented control logic for controlling of motor movements for complex motor such as Brushless Direct Current (BLDC) motor and Permanent Magnet Synchronous Motor (PMSM). Chapter 6 discusses in detail the usage of signed and unsigned, with descriptions of the consequences of mixing signed and unsigned in the RTL.

State machines are commonly found in ASIC and SOC design. Chapter 7 shows the different state machine encoding and RTL coding styles for designing state machine.

Blocking statements and non-blocking statements commonly found in procedural statements are also discussed in detail in Chap. 7.

Chapter 8 discusses common RTL coding guidelines that designers can follow to ensure good RTL code. Complete sensitivity list, edge-triggered RTL, level-triggered RTL, handling of bidirectional ports in RTL, when to use blocking statement and when to use non-blocking statement, inferred latch, signed Verilog, logic between modules, registering output of modules, and naming convention are discussed in detail in Chap. 8.

With today's complex large ASIC and SOC, it is important to ensure the RTL is well covered and verified with the testbenches. Code coverage is a method to measure the percentage of RTL code coverage by the corresponding testbenches. RTL code that are not exercised by the testbenches are identified by code coverage. Chapter 9 discusses about code coverage and shows the different types of code coverage. Chapter 9 also shows how designers can use the information obtained from code coverage reports to enhance the testbenches to cover the missing coverage.

All the chapters in this book consist of many pictures and diagrams to help the reader visualize and understand the design examples. Many examples are provided throughout the book to help assist the reader to understand the complications and issues presented and solutions to overcome them.

This book's target audience are students learning VLSI design and young design engineers learning the art of VLSI design. This book explains design from a practical perspective based on years of design knowledge gained from experience.

Bayan Lepas, Malaysia Weng Fook Lee

Trademarks

Pyxis, Calibre, Kronos, Modelsim, Leonardo Spectrum, DFT Advisor, Fastscan, Testkompress, Logic BIST, MBISTArchitect, and HDL Designer are trademarks of Mentor Graphics Inc.

Composer is a trademark of Cadence Inc.

Cyclone and Aria are trademarks of Altera/Intel Corp.

Acknowledgment

This book would not have been possible without the help of many people. I would like to put forward a word of "Thank You" to my wife for her companionship; I am looking forward to many more years of companionship on this journey of life; to my kids, for sitting with me at the study table while I write this book and they work on their school assignments; and to my besties, LP, SC, and BH, thank you for your patience in listening to my endless stories and for always being there for me.

Contents

List of Figures

List of Tables

List of Examples

Chapter 1
Introduction

In today's growth of mobile smartphones, tablets, lightweight laptops, and ultrabooks, complex and powerful IC chips are used extensively in almost every aspect of our daily lives. With smartphones reaching one billion units a year, complex design of System on Chip (SoC) and application-specific integrated circuit (ASIC) are driven with more features and capabilities. The increased functionality and features increase design complexity at a scale unseen before in the field of very-large-scale integration (VLSI) design.

Design of ASIC and SoC with millions of transistors poses a difficult challenge in VLSI design. Some complex advanced ASIC or SoC used in gaming console surpass one billion transistor count. Verification of such chips becomes a difficult task, with design bugs needed to be caught prior to design tapeout.

Fabrication cost of VLSI chips is very high; any design bugs caught in silicon after fabrication would require design fixes and fabrication respin, increasing the financial cost to bring the chip to the market.

Design of ASIC and SoC is complex and difficult. Verification of multimillion transistor design is equally as complex. Effective and efficient design methodology is needed to aid the design of ASIC and SoC. This book shares the design knowledge gained by experienced designers, covering design methodology, multi-clock design, latch inference, scan chain, ATPG, logic BIST, memory BIST, signed Verilog, state machine, common RTL coding rules, and code coverage.

W. F. Lee, *Learning from VLSI Design Experience*,
https://doi.org/10.1007/978-3-030-03238-8_1

Chapter 2
Design Methodology and Flow

All integrated circuits (IC) are designed using a predefined flow and methodology. This flow and methodology allow the design engineer to follow each step of the flow and to take the appropriate action along the flow to address any problems that arise.

Design methodology and flow can be categorized into three distinct flow, analog/custom design flow, digital design flow, and mixed signal design flow. Each flow is catered to its particular design, such as analog/custom design flow for analog/custom circuits. Similarly digital design flow is used for digital design and mixed signal design flow for mixed signal design.

The design flow and methodology shown in Fig. 2.1 are a generic analog/custom design flow, with Fig. 2.7 showing a generic digital design flow. For mixed-signal design which consists of some blocks analog/custom and some blocks digital, the flow is a combination of both the analog/custom design flow of Fig. 2.1 and the digital design flow of Fig. 2.7. However, some design center may make some changes in certain parts of the flow to cater to their design needs.

Analog/Custom Design Flow

As its name implies, analog/custom design flow is used for design of analog or custom circuits. Custom circuits refer to design of circuits using components of transistors, resistors, and capacitors and may not necessarily be analog.

1. Schematic
   (a) This part of the process is also referred to as schematic capture.
   (b) An analog/custom designer uses computer-aided design (CAD) software tools such as Mentor Graphics' Pyxis Schematic Capture or Cadence's Composer to draw the schematic circuit of an analog or custom design.

W. F. Lee, *Learning from VLSI Design Experience*,
https://doi.org/10.1007/978-3-030-03238-8_2

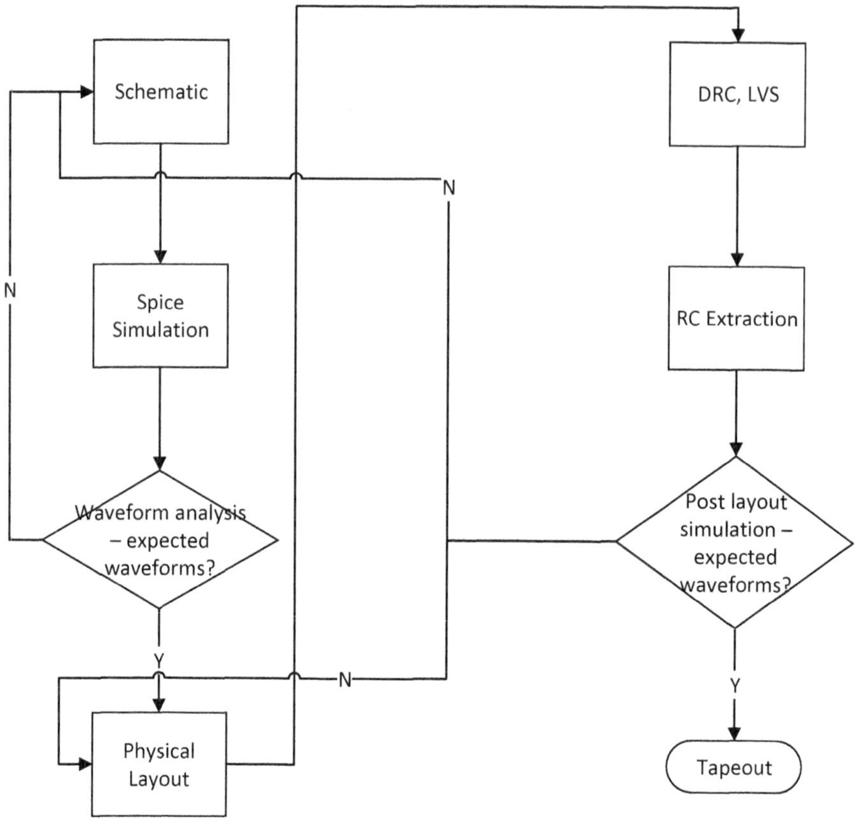

Fig. 2.1 Diagram showing analog/custom design flow

(c) The schematic can compose of transistors, gates, resistors, capacitors, and inductors, basically components that form the functionality of the analog/custom circuit.

(d) Figure 2.2 shows an example of a schematic of an inverter using Mentor Graphics' Pyxis Schematic Capture tool.

2. Spice simulation

   (a) This part of the process involves pumping of stimulus into the circuit, and the behavior of the circuit is analyzed to determine if the circuit behaves as expected.

   (b) To enable simulation of the circuit, a testbench is needed. It is a setup for appropriate stimulus to be pumped into design under test (DUT).

   (c) Figure 2.3 shows a testbench to drive the VDD at 5 V with reference to VSS and a pulse of 5 V with a width of 20 ns into input of the inverter.

3. Waveform analysis
   (a) The behavior of the circuit as observed in simulation is analyzed to determine if the circuit is behaving as expected.
   (b) Referring to Fig. 2.4, if the behavior of the DUT is not as expected, the designer will go to the schematic capture phase or Testbench phase to fix the bug and resimulate. This will progress in a loop until the designer is satisfied that the DUT is behaving as expected.
4. Physical layout
   (a) This part of the process is the translation from schematics to physical layers in layout that can be fabricated to form the chip.
   (b) Different physical layers such as metal, poly, active layer, and others are drawn to form the physical representation of the design (Fig. 2.5).
5. DRC, LVS
   (a) Design rule check or DRC is a process in which the physical layout which consists of polygons and squares of different layout layers is checked for fabrication design rules. Some examples are minimum metal width and minimum active length. The more technologically advanced the fabrication process, the more design rules that must be checked. For example, a 45 nm fabrication process will have many more complex design rules to check as compared to a 0.13 μm fabrication process.
   (b) Typically, a DRC process is verified using a DRC tool such as Mentor Graphics' Calibre DRC. When any DRC error is reported, the layout designer must change the layout at those locations where error is highlighted. This is repeated until all DRC errors are fixed.
   (c) After DRC is verified clean, the flow proceeds to layout versus schematic (LVS). This process verifies that the physical layout matches schematic. This part of the flow is needed as simulations on a circuit's functionality are verified on the schematic; therefore, the physical layout has to be verified to ensure it matches schematics. If any mismatches are caught during LVS, the error must be fixed in physical layout. This is repeated until all the LVS errors are fixed (Fig. 2.6).
6. RC extraction
   (a) This is an important flow after physical layout is completed. In this part of the flow, the interconnects in physical layout are extracted for its resistance and capacitance (RC) together with the circuits' transistors and its corresponding RC. This process creates a spice netlist that contains all the circuits' components and RC values. This is referred to as post layout netlist.
   (b) Post layout netlist consists of resistors, capacitors, and inductors that do not exist in schematics as these components originate from layout routings and component parasitics.

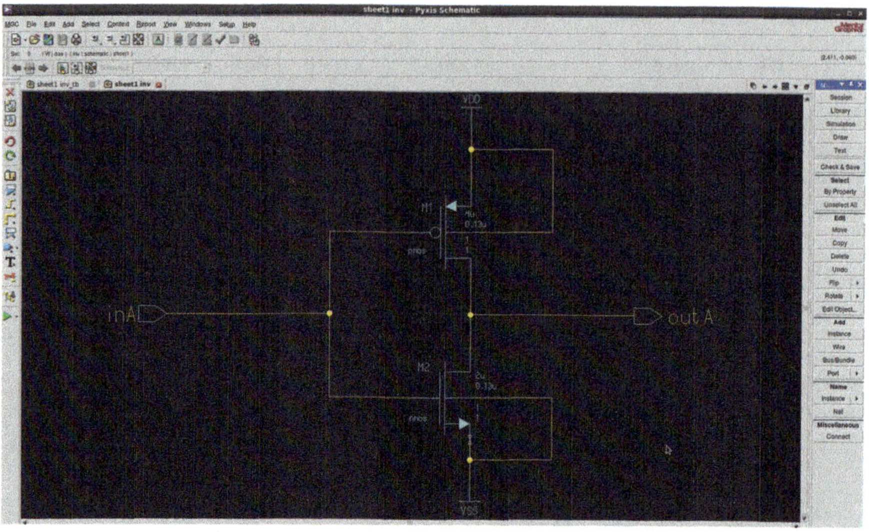

Fig. 2.2 Diagram showing schematic of an inverter

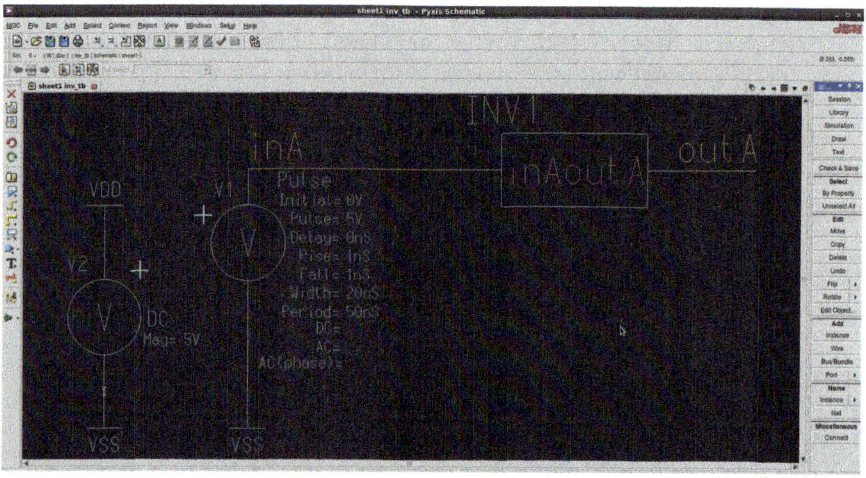

Fig. 2.3 Diagram showing schematic of testbench to pump stimulus

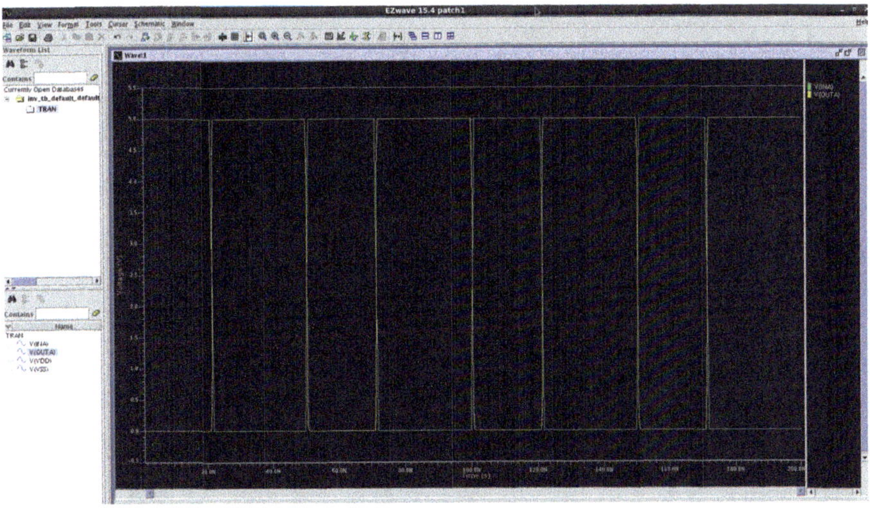

Fig. 2.4 Diagram showing waveform from simulation

7. Post layout simulation
   (a) This is the final part of the flow prior to tapeout.
   (b) The extracted netlist is resimulated using the same test vectors and the waveform observed and analyzed to determine if the circuit is behaving as expected.
   (c) There are occasions where the extracted netlist does not behave as expected due to the RC parasitic. The designer will have to make modifications to the circuit either at the layout level or schematic level, and the flow repeats itself again. This is repeated until the post layout netlist passes simulation, at which point the database is ready for tapeout.
   (d) When the post layout extracted netlist passes simulation, the GDSII file format for the physical layout is generated and passed to a fab for fabrication.

Digital Design Flow (Fig. 2.7)

Digital design flow is used for digital circuits and typically used when the digital circuits are design using hardware description language (HDL).

1. RTL Coding
   (a) Register-transfer level (RTL) coding is the process in which a designer describes the functionality of a digital circuit using HDL language.
   (b) HDL language consists of either Verilog HDL or Very High Speed Integrated Circuit Hardware Description Language (VHDL).
   (c) Figure 2.8 shows an example of a RTL Verilog code.

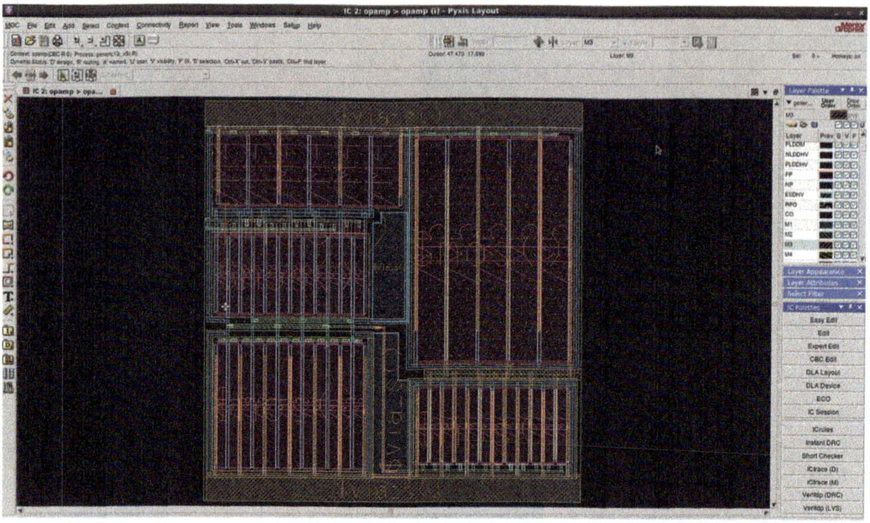

Fig. 2.5 Diagram showing physical layout of a circuit

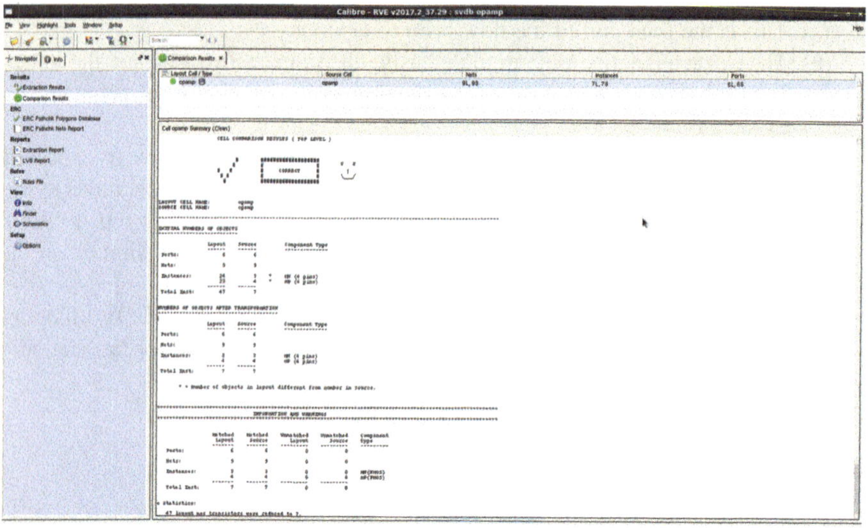

Fig. 2.6 Diagram showing a circuit verified clean on DRC and LVS

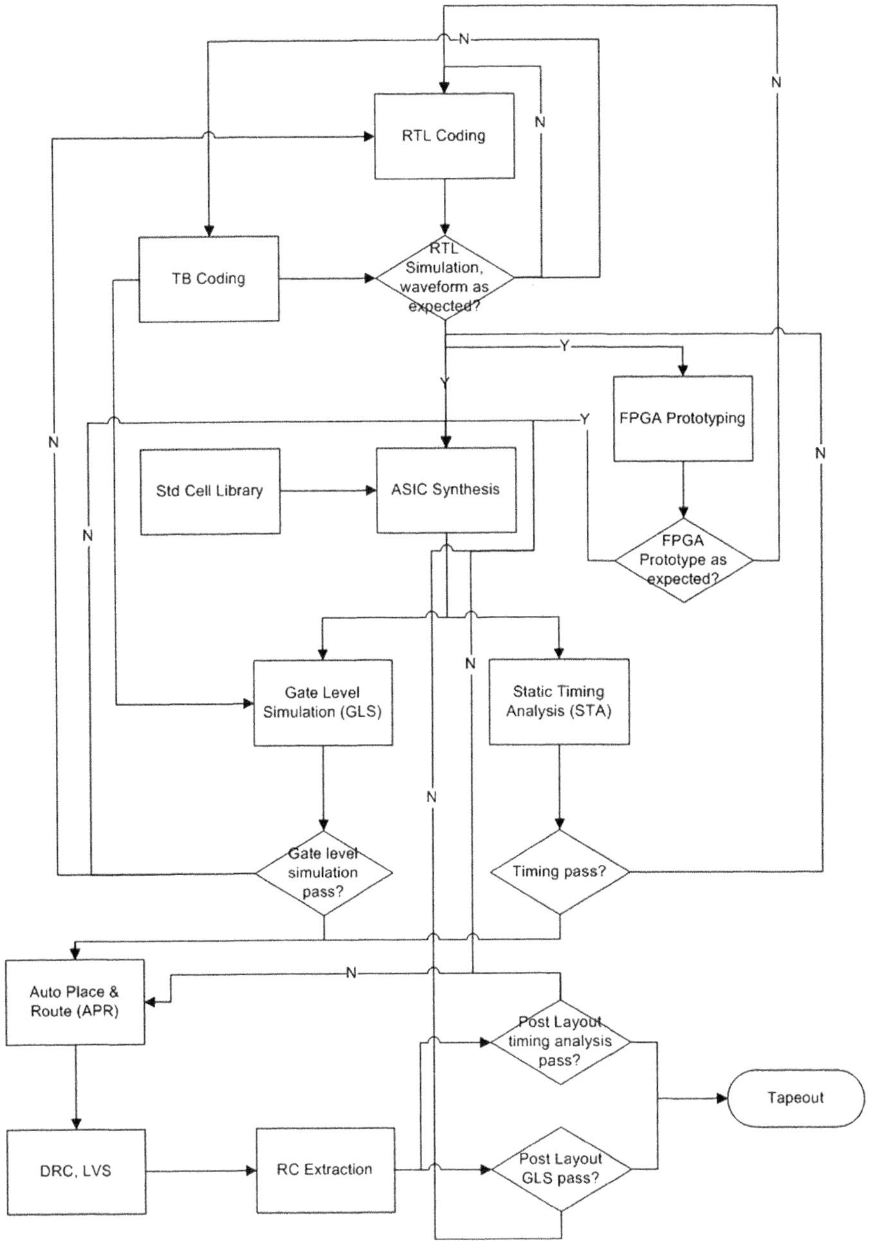

Fig. 2.7 Diagram showing digital design flow

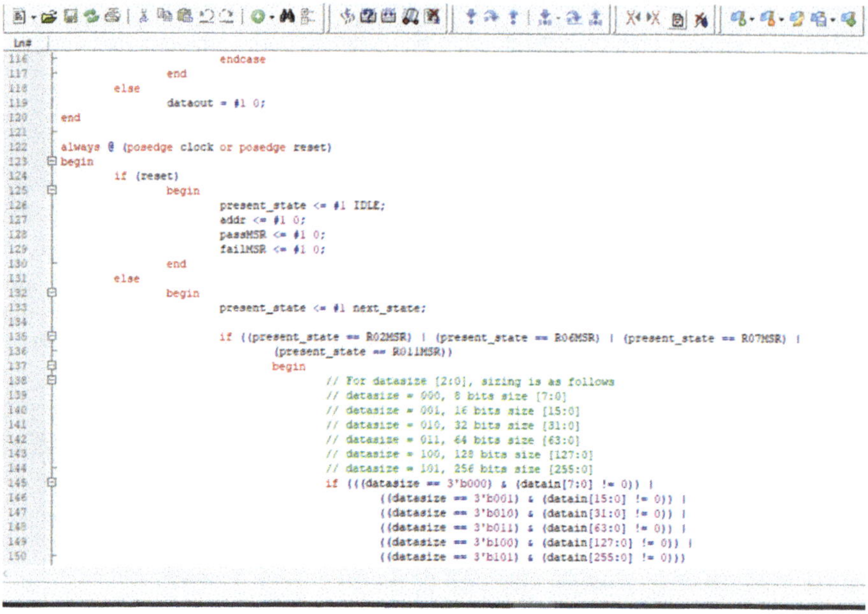

Fig. 2.8 Diagram showing example of a RTL Verilog code

2. Testbench Coding

   (a) Testbench coding is the process in which the designer writes a HDL code to pump in stimulus into the DUT and observe the output of the RTL code.
   (b) Testbench is commonly coded in HDL, either Verilog or VHDL. Some testbenches may use System Verilog.
   (c) Figure 2.9 shows an example of a Verilog testbench to pump stimulus into a design.
   (d) Although both RTL code and testbench are written in Verilog or VHDL, there is a major difference between them. RTL code is targeted for synthesis, while testbench is only for simulation and verification of the design.

3. RTL Simulation

   (a) In RTL simulation, a simulator such as Mentor Graphics' Modelsim can be used to compile the testbench and RTL code.
   (b) In simulation, the testbench will pump the specified stimulus into the RTL design.
   (c) The designer can plot the waveforms of signals in the RTL and observe the behavior of the RTL to determine if it is behaving as expected.
   (d) If a bug in the RTL code is found and the RTL code does not function as expected, the fix must be made in the RTL, and the simulation process repeats until the designer is satisfied that the RTL code is functioning correctly.

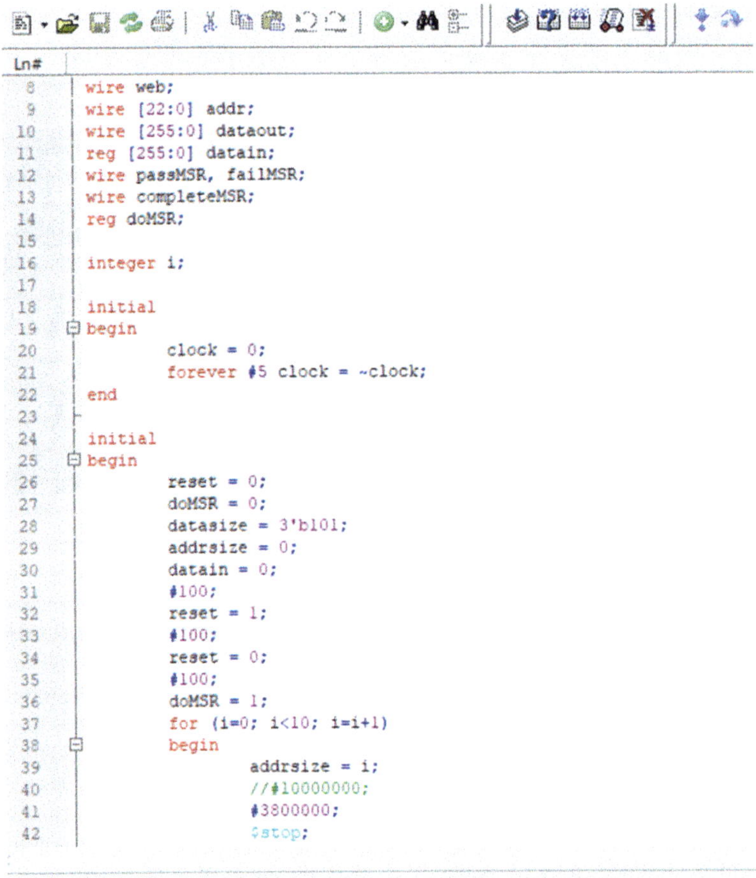

Fig. 2.9 Diagram showing example of a Verilog testbench code

(e) A bug may also occur in the testbench code rendering the testbench to pump in wrong stimulus. The bug in the testbench must be fixed and resimulated. This is repeated until all the bugs in the testbench are caught and fixed.

(f) Figure 2.10 shows an example of a simulation waveform. The designer will analyze the waveform and determine if the RTL simulation behaves as expected. If it is not and bugs are found, the bugs need to be fixed and simulation rerun. When the simulation waveform behaves as expected, the designer proceed to the next step which is FPGA prototyping.

4. FPGA Prototyping

(a) Field-programmable gate array can be used by digital designers to prototype the digital circuits.

(b) The designer can synthesize their RTL code into a particular FPGA technology such as Altera's Cyclone series or Aria series and then download the

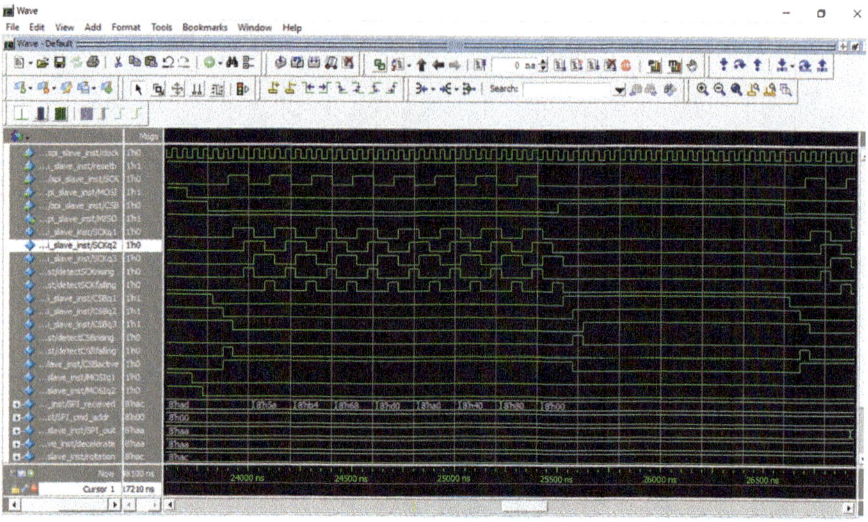

Fig. 2.10 Diagram showing example of a simulation waveform

specific bit file into the FPGA. The FPGA will behave with the functionality of the synthesized RTL code.

(c) The designer can proceed to test the FPGA as the designer's digital circuit.

(d) There are many advantages to FPGA prototyping:

A digital design can be prototyped and tested even before the design is fabricated.

As cost of fabrication is high, any bugs in the digital design can be caught at the FPGA level, and the bug fixed in RTL. The designer can resynthesize the RTL fix and reprogram the FPGA. The design can be retested for functionality.

When a designer wants to fabricate a digital design, the designer needs to have large volume. Each fabrication may have several hundreds of die on a single wafer. However, for prototyping, the designer only intends to test the digital circuit; therefore, large volume is not required. A few FPGA chips are adequate for prototyping. FPGA chips are sold in small volume thereby allowing the designer to purchase only several units and prototype the digital circuit.

FPGA allows the digital designer a low-cost means of prototyping and testing their digital circuit without the need for fabrication.

(e) FPGA prototyping however also carries several disadvantages:

Digital design implemented in FPGA is typically slower than a fabricated chip. Therefore, digital design implemented in FPGA is tested for its functionality, not its timing.

If a digital designer needs a large volume of chips for the design, it is not practical to use FPGA to implement the design as FPGA per unit cost is high for large volume. For example, in application-specific integrated circuit (ASIC) fabrication using a matured technology such as 0.18 μm, $150,000 may fabricate 100,000 die depending on circuit size, thereby having an average of $2.50 per chip after adding in $1.00 for packaging cost and testing cost. However, cost per FPGA typically range much higher. For large volume, ASIC fabrication is more suitable compared to FPGA.

5. Standard Cell Library

   (a) A standard cell library is a library needed for synthesis of digital circuits.
   (b) It consists of many different flavors of different types of logic gates, ranging from basic logic gates such as inverters to complex logic gates.
   (c) Each logic gate type has different sizing for different drive strengths.
   (d) Examples of some logic gates in a standard cell library.

   inv_x (inverter)

   _x represents the sizing. A "b" has bigger drive strength than an "a." Similarly, a "c" has bigger drive strength than a "b."

   A bigger drive strength translates to the transistors in the logic gate having bigger W/L ratio.

   Referring to Fig. 2.11, the W/L of the pmos is 2× of the nmos. This is due to holes (pmos carrier) being 2× weaker than electrons (nmos carrier).

   From inv_a to inv_b to inv_c, the drive strength increases, allowing the bigger inverter to drive bigger loads.

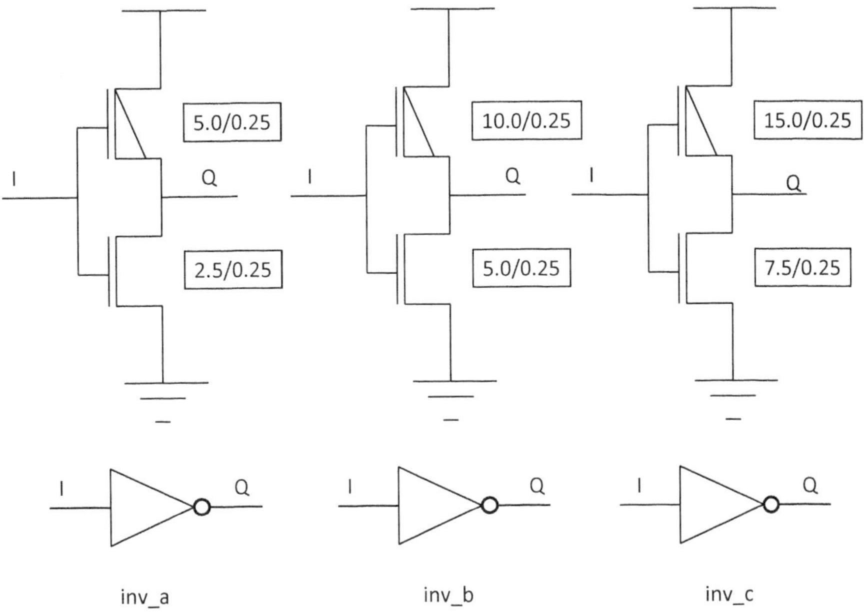

Fig. 2.11 Diagram showing example of different sizing for inverter in a standard cell library

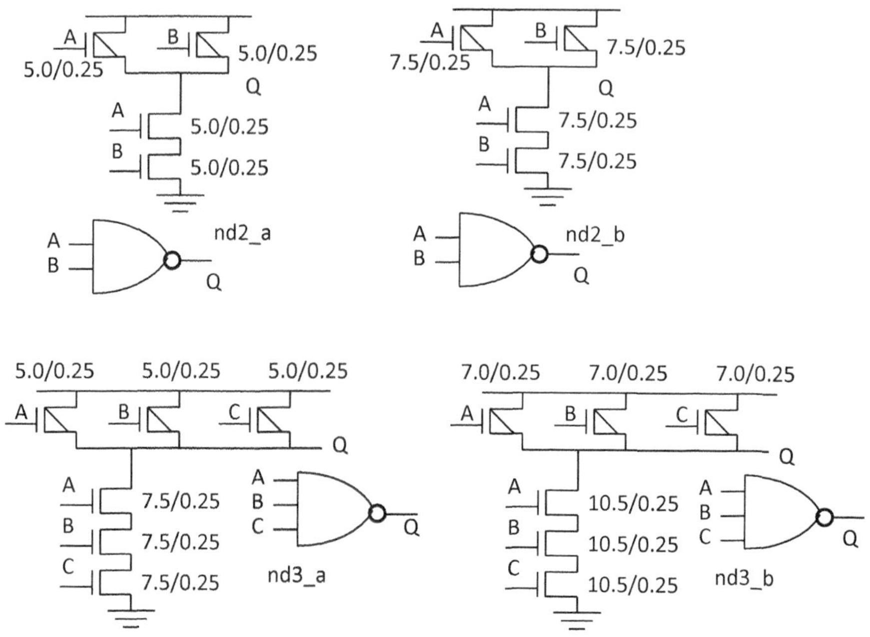

Fig. 2.12 Diagram showing example of different sizing for 2 input NAND and 3 input NAND in a standard cell library

During synthesis, if a gate drives a large fanout or a gate drives several larger gates, the synthesis tool will automatically upsize the gate to have a larger drive strength.

nd2_x, nd3_x, nd4_x

Referring to nd2_a in Fig. 2.12, the two nmos in series will weaken the nmos drive strength. The effective drive strength of the nmos transistors is 5/0.25 divided two, making it 2.5/0.25.

Since pmos are twice as weak as nmos, to have the same drive strength as the nmos, the pmos effective size is 5.0/2.5.

Referring to nd3_a in Fig. 2.12, the three nmos in series will cause the effective drive strength divided three. Each nmos is sized 7.5/0.25 to give an effective nmos drive strength of 2.5/0.25. The pmos remains at 5.0/0.25 since it is twice as weak as the nmos.

nr2_x, nr3_x

Referring to nr2_a in Fig. 2.13, the two pmos in series will weaken the pmos drive strength. The effective drive strength of the pmos transistors is 10/0.25 divided two, making it 5.0/0.25.

Since nmos are twice as strong as pmos, to have the same drive strength as the pmos, the nmos effective size is 2.5/0.25.

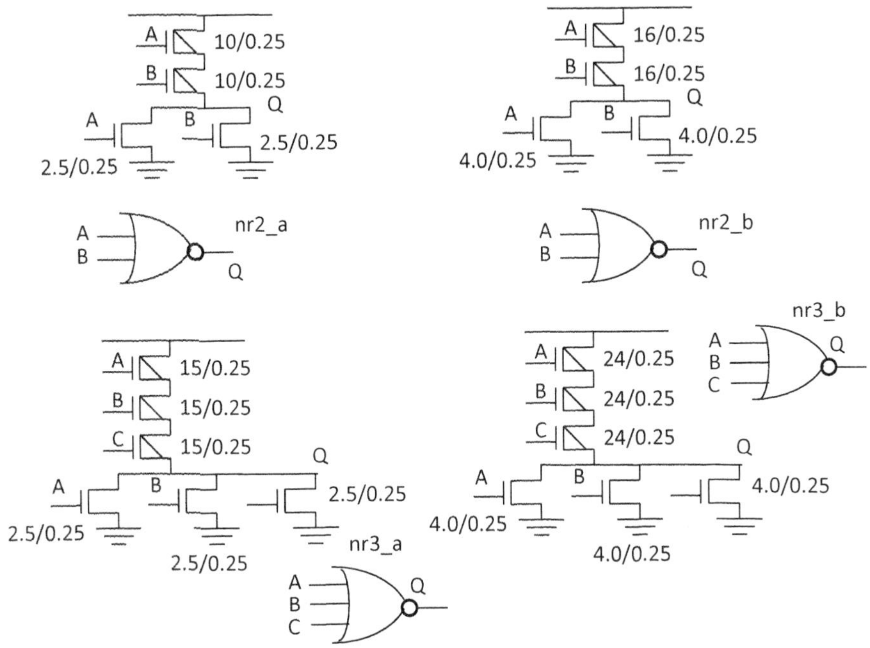

Fig. 2.13 Diagram showing example of different sizing for 2 input NOR and 3 input NOR in a standard cell library

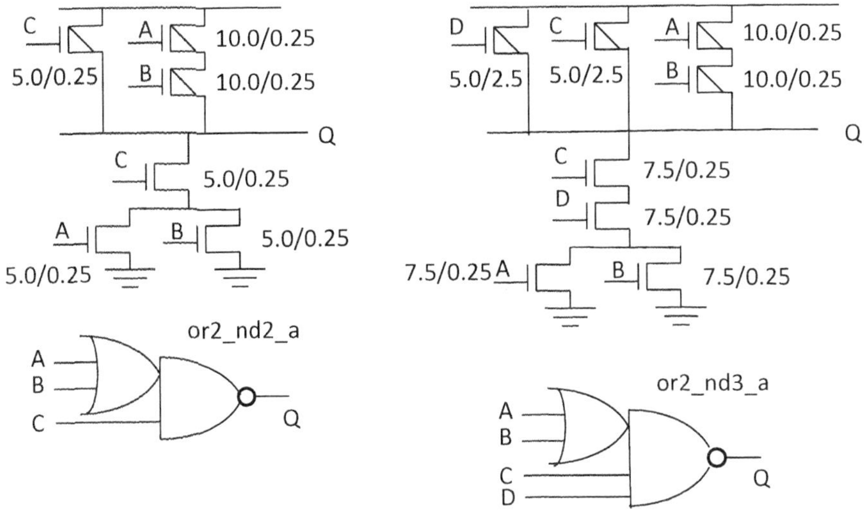

Fig. 2.14 Diagram showing example of 2 input OR with 2 input NAND (or2_nd2_x) and 2 input OR with 3 input NAND (or2_nd3_x) in a standard cell library

Referring to nr3_a in Fig. 2.13, the three pmos in series will cause the effective drive strength divided three. Therefore, each pmos is sized 15.0/0.25 to give an effective pmos drive strength of 5.0/0.25. The nmos is twice as strong as the pmos, thus its size of 2.5/0.25.

or2_nd2_x, or2_nd3_x

Referring to or2_nd2_a in Fig. 2.14, the two pmos in series will weaken the pmos drive strength. The effective drive strength of the pmos transistors is 5/0.25.

For the nmos that are in series, their effective drive strength is 2.5/0.25.

Referring to or2_nd3_a in Fig. 2.14, the three nmos transistors are in series; thus, the sizing of each transistor is divided three in order to obtain the effective drive strength of 2.5/0.25. And the two pmos in series have a size of 10/0.25 to achieve a drive strength of 5.0/0.25.

an2_nr2_x, an2_nr3_x

Referring to an2_nr2_a in Fig. 2.15, the two pmos in series will weaken the pmos drive strength. The effective drive strength of the pmos transistors is 5.0/0.25.

For the nmos that are in series, their effective drive strength is 2.5/0.25.

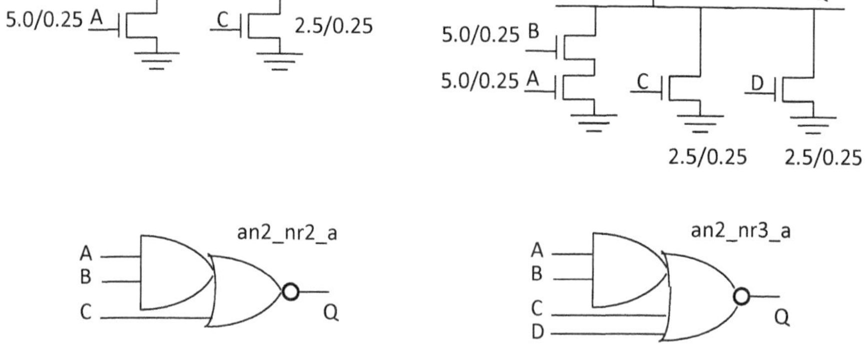

Fig. 2.15 Diagram showing example of 2 input AND with 2 input NOR (an2_nr2_x) and 2 Input AND with 3 input NOR (an2_nr3_x) in a standard cell library

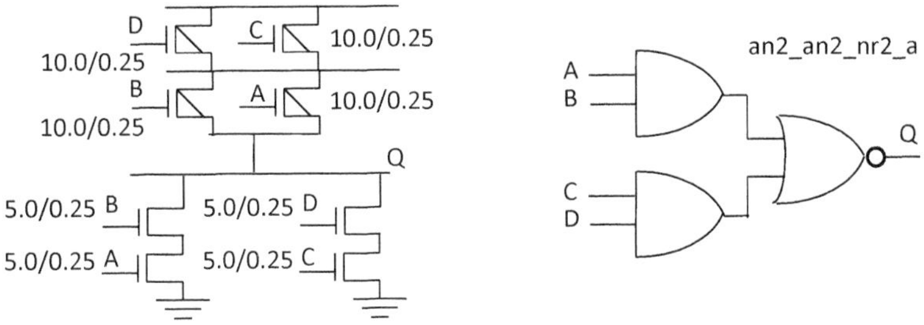

Fig. 2.16 Diagram showing example of two 2 input AND with 2 input NOR (an2_an2_nr2_x) in a standard cell library

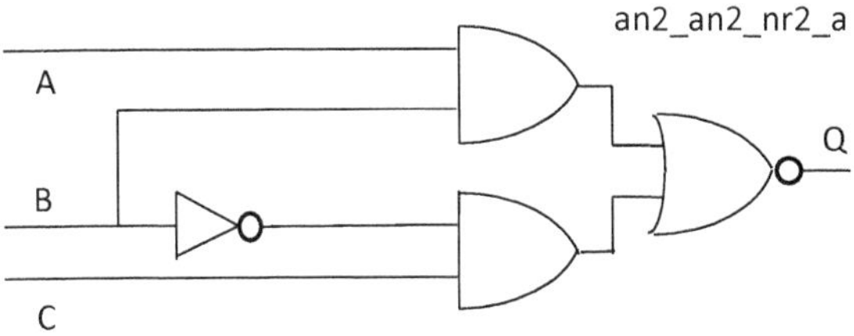

A	B	C	Q
0	0	0	1
0	0	1	0
0	1	0	1
0	1	1	1
1	0	0	1
1	0	1	0
1	1	0	0
1	1	1	0

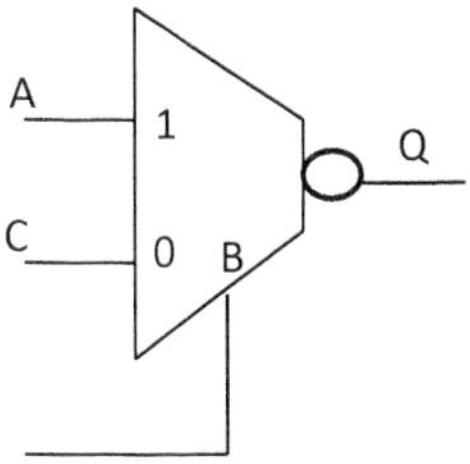

Fig. 2.17 Diagram showing usage of an an2_an2_nr2_x to form 2 to 1 inverted multiplexer

Referring to an2_nr3_a in Fig. 2.15, the three pmos transistors are in series; thus, the sizing of each transistor is divided three in order to obtain the effective drive strength of 5.0/0.25.

an2_an2_nr2_x

Referring to an2_an2_nr2_a in Fig. 2.16, this complex gate can typically be used together with an inverter to form a 2 to 1 inverted multiplexer as shown in Fig. 2.17.

Referring to Fig. 2.17, B acts as a selector to select either A or C. When B is 0, Q is inverted of C. When B is 1, Q is inverted of A.

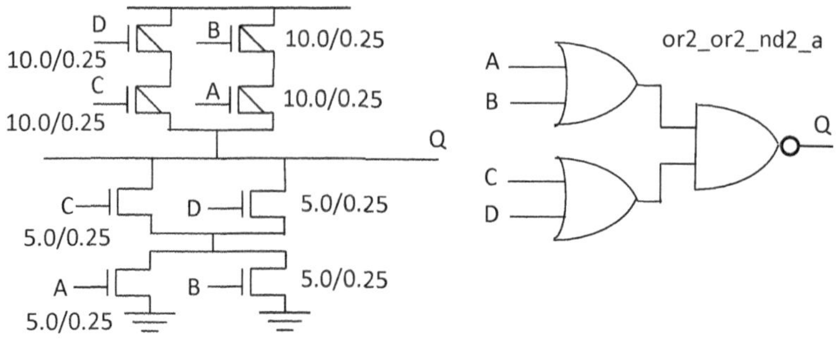

Fig. 2.18 Diagram showing example of two 2 input OR with 2 input NAND (or2_or2_nd2_x) in a standard cell library

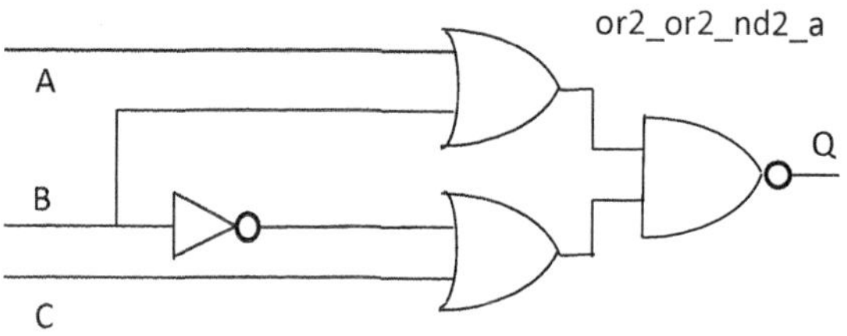

A	B	C	Q
0	0	0	1
0	0	1	1
0	1	0	1
0	1	1	0
1	0	0	0
1	0	1	0
1	1	0	1
1	1	1	0

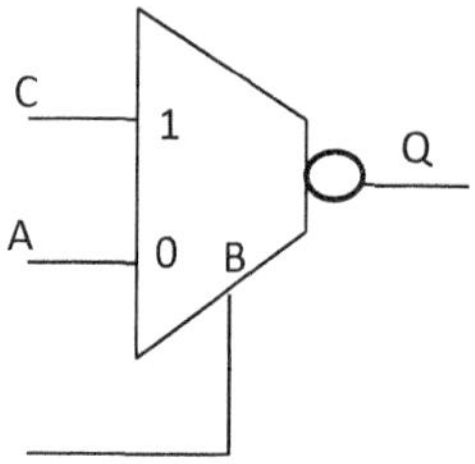

Fig. 2.19 Diagram showing usage of a or2_or2_nd2_x to form 2 to 1 inverted multiplexer

or2_or2_nd2_x

Referring to or2_or2_nd2_a in Fig. 2.18, this complex gate can be used together with an inverter to form a 2 to 1 inverted multiplexer as shown in Fig. 2.19.

Referring to Fig. 2.19, B acts as a selector to select either A or C. When B is 0, Q is inverted of A. When B is 1, Q is inverted of C.

Data latch

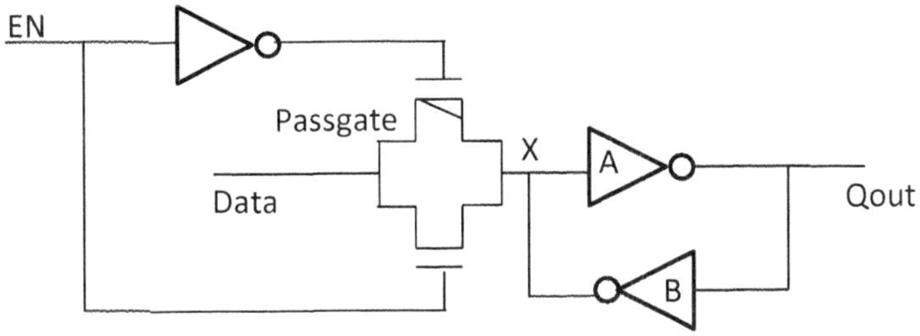

Fig. 2.20 Diagram showing A basic D latch

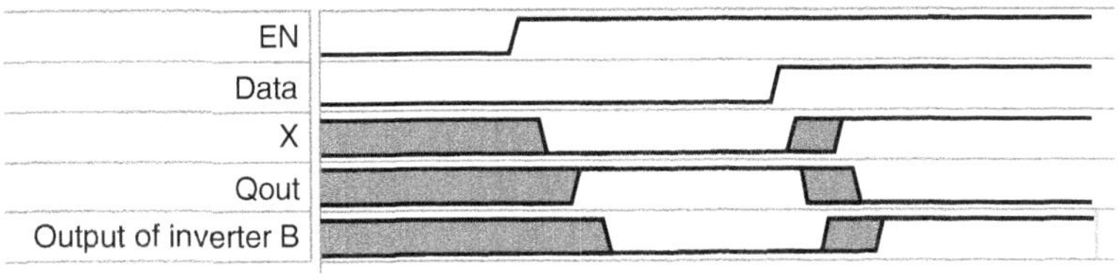

Fig. 2.21 Diagram showing contention occurring in Node X

Figure 2.20 shows a basic data latch. The sizing of the passgate must be stronger compared to inverter B.

If EN = 1, the pmos passgate and nmos passgate switches on, a contention may occur at node X. for example, assuming the data to the basic latch changes from 0 to 1, while EN is high as shown in Fig. 2.21, contention may occur.

Referring to Fig. 2.21, when EN goes high, the logic 0 from the data is driven onto node X, thereby giving node X a logic 0. Qout is inverted of node X thus giving a logic 1. And inverter B inverts Qout giving a logic 0 on the output of inverter B thus driving node X with logic 0.

However, when the data changes from 0 to 1, output of inverter B is still driving a logic 0 on node X, while the data passes through the passgate and driving node X as logic 1. This causes a contention, and a VCC to VSS short occurs. In order for the D latch circuit of Fig. 2.20 to function as a latch, inverter B must be a weak inverter thereby allowing the passgate to win the contention.

Inverter B can be a weak inverter by using weak pmos transistor and weak nmos transistor to form the inverter. To achieve a weak transistor, the effective W/L is reduced, with L being larger. The larger the L, the weaker the transistor. Figure 2.22 Shows the diagram of the D latch with the weak transistors used to form inverter B.

Referring to Fig. 2.22, note that the length (L) of the pmos and nmos transistors has a large L as compared to the normal inverter A.

The basis for the functionality of the D latch is the weak inverter B being "over-powered" by the strong passgate when the passgate is driving a different logic values as opposed to inverter B.

Reset latch

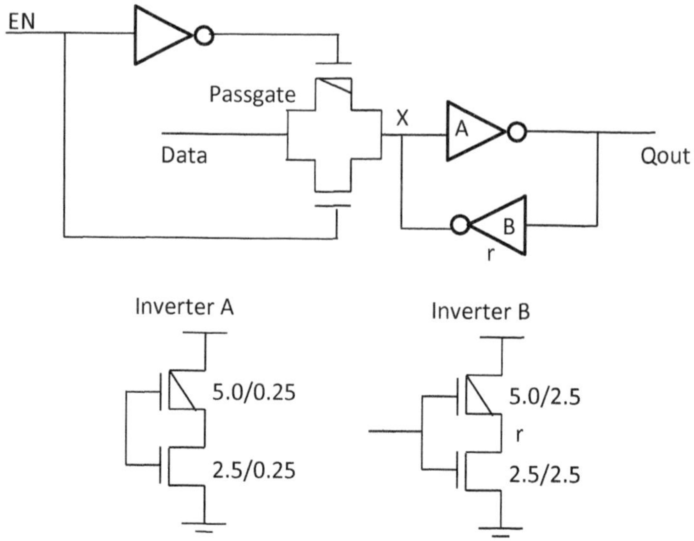

Fig. 2.22 Diagram showing a weak inverter B with large L

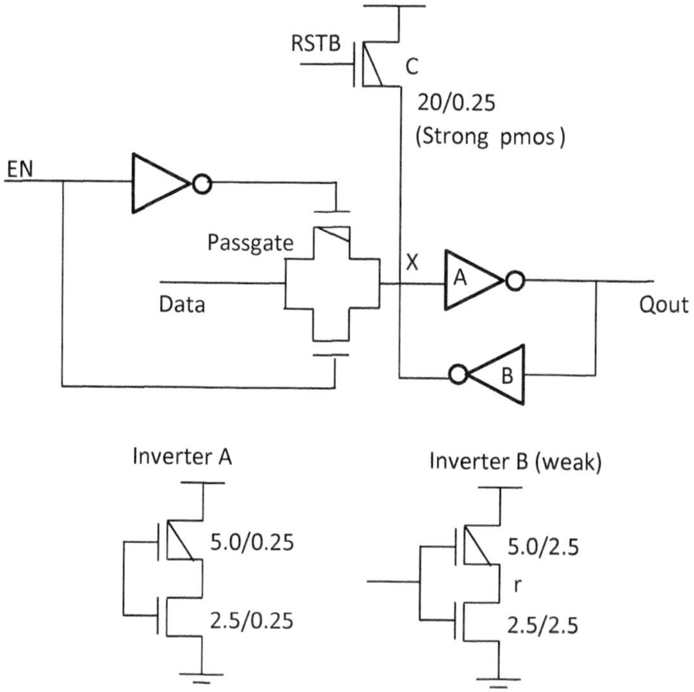

Fig. 2.23 Diagram showing reset latch with strong reset pmos

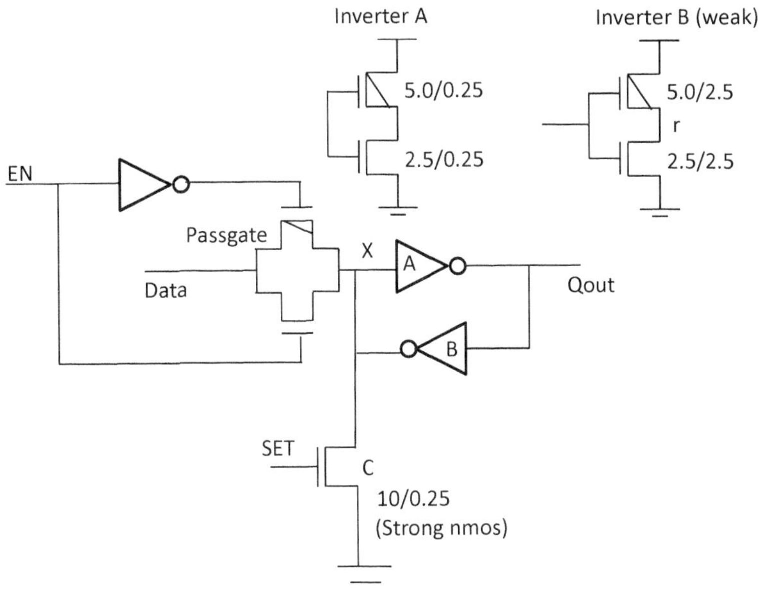

Fig. 2.24 Diagram showing set latch with strong set Nmos

Referring to Fig. 2.23, the RSTB pmos must be a strong pmos. When RSTB is at logic 0, the reset pmos is switched on to pull node X to a logic 1 resulting in Qout being a logic 0.

However, if the existing value at node X prior to RSTB going to logic 0 is a value of logic 0, a contention will occur when RSTB goes to logic 0. Node X will have a contention between weak logic 0 driven by the weak inverter B and strong logic 1 driven by the strong pmos C. As pmos C is a stronger pmos, it will win over the weak inverter B, thereby driving logic 1 to node X and Qout to logic 0 during a reset.

Pmos C must be stronger than the passgate to ensure that the reset always occur when RSTB is driven to logic 0 and pulls node X to logic 1.

Set latch

Referring to Fig. 2.24, the SET nmos must be a strong nmos. When SET is at logic 1, the set nmos is switched on to pull node X to a logic 0 resulting in Qout being a logic 1.

However, if the existing value at node X prior to SET going to logic 1 is a value of logic 1, contention will occur. When SET goes to logic 1, node X will have a contention between weak logic 1 driven by the weak inverter B and strong logic 0 driven by the strong nmos C. As nmos C is a stronger nmos, it will win over the weak inverter B, thereby driving logic 0 to node X and Qout to logic 1 during a set.

Nmos C must be stronger than the passgate to ensure node X is always pulled down to logic 0 when SET is logic 1.

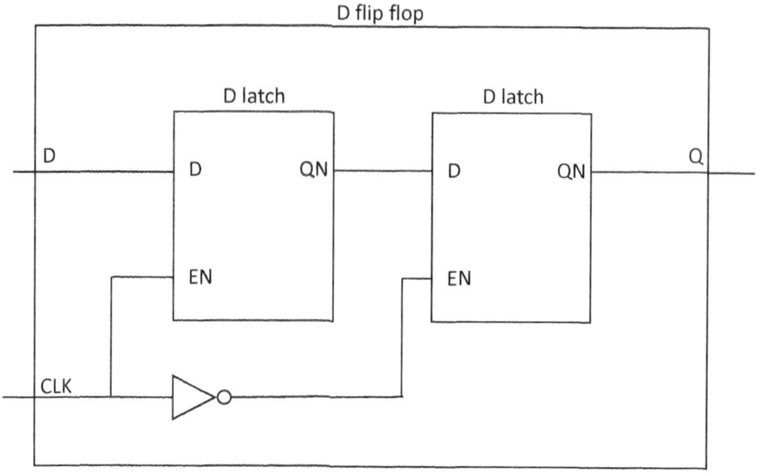

Fig. 2.25 Diagram showing a basic data flip-flop using master slave latch

D Flip-flop (using master slave latch).

Referring to Fig. 2.25, a basic data flip-flop can be built using two data latches back to back with an inverter inverting the clock signal between both the latches.

The schematic shown in Fig. 2.25 is commonly referred to as master slave data flip-flop.

The data path timing from D of the master latch to D of the slave latch must be longer than the clock path through the inverter; otherwise, a feed through may occur.

(e) Standard cell library is a library used in logic synthesis in which RTL code is converted into logic gates by synthesis tools. The synthesized logic circuit depends on the cells available in a standard cell library. Typically, a standard cell library with more cells will allow the synthesis tool to have better logic optimization. However, large standard cell library would require a lot of resources to characterize the cells to generate the .lib file. Standard cell libraries are in liberty format (.lib). The .lib file have all the characterization information about the cells available in the library, such as rise time, fall time, and transition time based on certain load. A standard cell library typically has three .lib file, one for fast corner (typically named min), one for slow corner (typically named max), and one for typical corner (typically named typ).

6. ASIC synthesis

(a) In synthesis, the functionality of the design written in HDL (Verilog or VHDL) is synthesized into logic gates. The synthesis tool will read the RTL code and converts or synthesizes it into logic circuits using the gates available in the standard cell library.

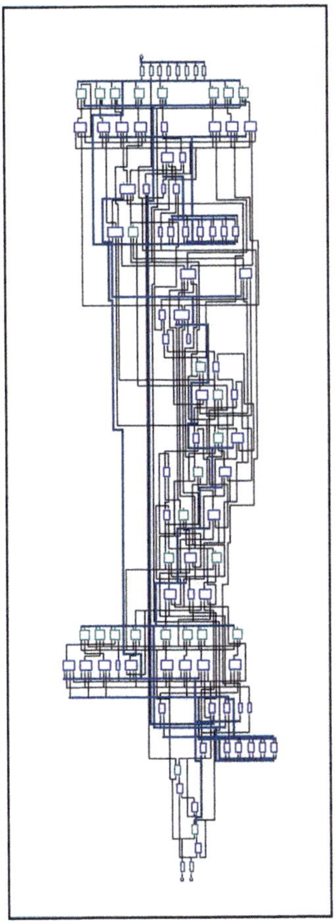

Fig. 2.26 Diagram showing schematic of a synthesized circuit

(b) A standard cell library with more flavors will allow synthesis tool to achieve better optimization. However, larger standard cell library will require more resources to characterize the cells into the .lib file.
(c) During synthesis, the designer inputs the RTL, standard cell library (.lib), and the timing constraints.
(d) Timing constraints are referred to as timing requirements for the design being synthesized. Examples are clock frequency, input delay, output delay, and combinational delay.
(e) Based on the logic cells available in the standard cell library, the timing constraints, and the RTL code, the synthesis tool will synthesize and optimize the logic circuits for optimal results.
(f) Figure 2.26 shows the schematic of a synthesized logic circuit.
(g) The output from the synthesis tool is a gate level netlist of the design synthesized into logic gates. It is commonly in Verilog format; however, VHDL gate level netlist can also be used.
(h) Referring to Example 2.1, the Verilog gate level netlist shows logic gates from the standard cell library being interconnected to form the logic design. The gate level netlist is output from synthesis tool and input to subsequent gate level simulation, static timing analysis, and auto place & route.

Example 2.1 Example showing a Gate Level Verilog Netlist Generated from Mentor Graphics' Leonardo Spectrum ASIC Synthesis

```
module fibgen ( clock, reset, fibout ) ;
    input clock ;
    input reset ;
    output [7:0]fibout ;
      wire int_clock, int_reset, int_fibout_7, int_fibout_6, int_
fibout_5, int_fibout_4, int_fibout_3, int_fibout_2, int_fibout_1, int_
fibout_0, sum_7, sum_6, sum_5, sum_4, sum_3, sum_2, sum_1, sum_0,
B_7, B_6, B_5, B_4, B_3, B_2, B_1, B_0, A_7, A_6, A_5, A_4, A_3,
A_2, A_1, A_0, sum_add_18_blk_carry_1, sum_add_18_c_1, sum_
add_18_c_1__dup_36, acc_A_op_r_6, acc_A_op_r_5, acc_A_op_r_4,
acc_A_op_r_3, acc_A_op_r_2;
     FA2A0 sum_add_18_DO_IO (.S1 (sum_0), .S2 (sum_1), .CO (sum_
add_18_c_1), .A1 (A_0), .A2 (A_1), .B1 (B_0), .B2 (B_1), .CI (gnd))
;
     FD1H0 reg_fibout_0 (.Q (int_fibout_0), .XQ (nx231), .D (nx44),
.CK (int_clock), .EN (NOT_FSM_current_state_0)) ;
    AO2A0 ix45 (.X (nx44), .A1 (FSM_current_state_0), .A2 (nx231),
.B1 (sum_0), .B2 (nx36)) ;
     FD1A2 FSM_reg_current_state_0 (.Q (FSM_current_state_0), .XQ
(NOT_FSM_current_state_0), .D (nx4), .CK (int_clock)) ;
     NR2R0 ix5 (.X (nx4), .A1 (int_reset), .A2 (FSM_current_
state_0)) ;
```

```
// more Verilog code not shown //
    NR2R0 ix25 (.X (nx24), .A1 (int_reset), .A2 (nx18)) ;
    FD1A0 FSM_reg_current_state_1 (.Q (FSM_current_state_1), .XQ
(nx239), .D (nx24), .CK (int_clock)) ;
    AO2A0 ix183 (.X (nx182), .A1 (int_fibout_1), .A2 (nx142), .B1
(acc_A_op_r_1), .B2 (nx255)) ;
    AO2A0 ix239 (.X (nx238), .A1 (int_fibout_3), .A2 (nx142), .B1
(acc_A_op_r_3)
          , .B2 (nx255)) ;
endmodule
```

7. Gate Level Simulation
   (a) In this step, the Verilog gate level netlist from the synthesis tool is resimulated using the same testbench that was used to simulate the RTL Verilog code.
   (b) The gate level simulation is to verify the synthesized netlist have the correct functionality of the RTL code.
   (c) If the gate level simulation fails, the functionality of the gate level netlist does not match the RTL Verilog code.
   (d) Referring to Fig. 2.27, when gate level simulation fails, fixes must be made to the RTL and the process repeats.
   (e) The flow can only proceed to static timing analysis when gate level netlist passes gate level simulation. This occurs when gate level netlist matches the functionality of RTL.
8. Static Timing Analysis
   (a) Static timing analysis is the process in which timing for every path within the gate level netlist is calculated and determined if a setup violation or hold violation have occurred.
   (b) If a timing violation occurred, the synthesis process must be reoptimized to meet the timing violation or the design RTL modified and resynthesized to meet timing.
   (c) During static timing analysis, timing constraints are needed. The timing constraints specify the clock period, input delay, output delay, multicycle path, false path, and combinational logic path delay.
9. Auto, Place & Route
   (a) Auto place & route can be considered the equivalent of synthesis but for backend.
   (b) It does floor planning, placement, and routing for the gate level netlist that have passed static timing analysis.
   (c) The auto place & route process will generate the layout gate level netlist and GDSII layout netlist.
   (d) The auto place & route process is similar to the synthesis process, and it uses a physical layout standard cell library.

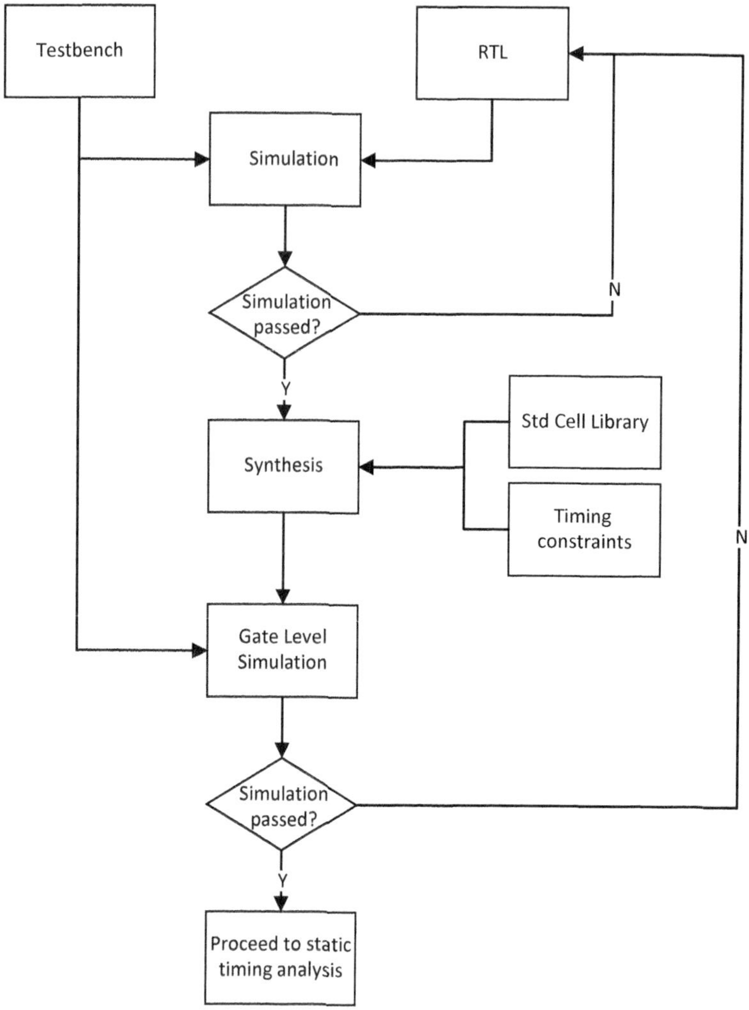

Fig. 2.27 Diagram showing process flow for gate level simulation

Each cell within the standard cell library has its physical layout.
During auto place & route, the corresponding cells in the gate level netlist will be placed using its physical layout cell in the standard cell library.
The physical layout cells of the gate level netlist will be placed in rows to share its VSS and GND as shown in Fig. 2.28.
The rows of cells are then routed to form the complete layout as shown in Fig. 2.29.

10. DRC, LVS

   (a) The GDSII layout netlist together with the layout database must be verified clean for design rule check (DRC) and layout versus schematic (LVS).

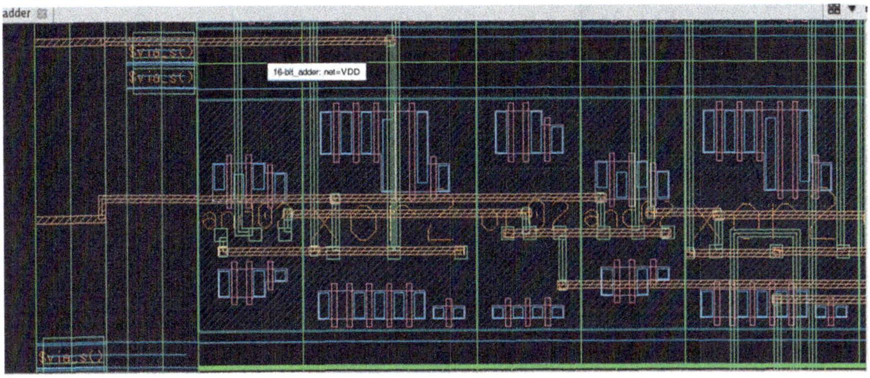

Fig. 2.28 Diagram showing physical layout with cells placed in rows to share VCC and GND

Fig. 2.29 Diagram showing physical layout of A 16 bit adder

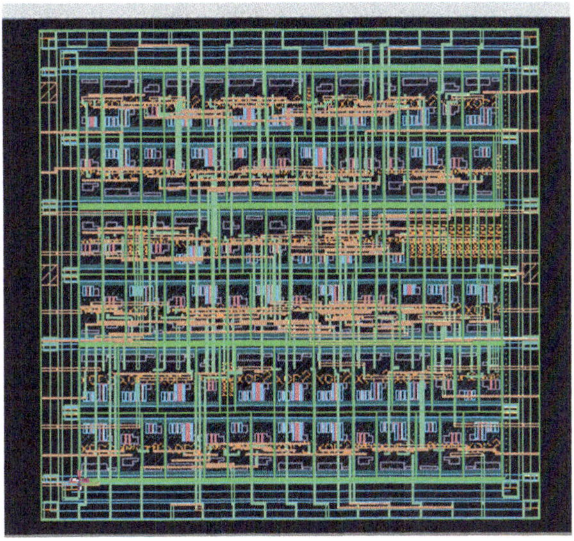

(b) Most fabs require design to be Calibre clean before it can fabricate the design. Calibre is Mentor Graphics Calibre series of tools which is the defacto standard used for DRC and LVS.

(c) In DRC, fabrication rules are checked to ensure the physical layout of the design is within the geometric rules of the fabrication process. Geometry rules such as metal width, metal spacing, active spacing, well spacing, minimum transistor channel length, and others are checked.

(d) Any DRC violation must be fixed in physical layout, and DRC reruns to ensure all DRC violations are fixed.

(e) In LVS, Calibre LVS analyzes the physical layout database and recognizes the corresponding transistors and interconnects in physical layout.

It compares the physical layout against schematic or gate level netlist to ensure they match.

(f) Any mismatch in LVS must be fixed in physical layout and the LVS process rerun.

11. RC Extraction

    (a) When physical layout have a clean DRC and LVS, the RC of the physical layout is extracted to obtain the interconnect as well as the transistors' equivalent R and C. The RC can be extracted in the form of RC in spice netlist or sdf (standard delay format) which is a standard format for specifying the interconnect delay as well as cell delay.
    (b) The RC extraction process to obtain the sdf is important as it allows the accurate timing from physical layout to be back annotated for post layout simulation and post layout static timing. The post layout analysis ensures the design works with the additional timing caused by the interconnects and physical layout of the cells.

12. Post Layout GLS

    (a) Post layout gate level simulation is similar to gate level simulation except that the netlist used for verification is physical layout gate level netlist instead of synthesized netlist.
    (b) In post layout GLS, accurate timing information from the RC extraction is back annotated to the post layout GLS simulation.
    (c) If a design passes the gate level simulation and static timing analysis, the design may fail in first pass post layout GLS as accurate timing information from physical layout database is back annotated to each net and each cell.
    (d) Any failure in post layout GLS will need to be fixed by modifing layout to minimize the timing path of the corresponding functionality failure.
    (e) If the failure cannot be fixed by modifying physical layout, post layout synthesis optimization is required.

13. Post Layout STA

    (a) Post layout STA is similar to static timing analysis except the netlist is post layout gate level netlist instead of synthesized gate level netlist.
    (b) In post layout STA, accurate timing information from RC extraction is back annotated to the post layout STA.
    (c) If a design passes gate level simulation and static timing analysis, the design may fail in first pass post layout STA as accurate timing information from physical layout database is back annotated to each net and each cell.
    (d) Any failure in post layout STA will need to be fixed by modifying the layout to minimize the timing path of the corresponding setup failure or maximize the timing path of the corresponding hold failure.
    (e) If the failure cannot be fixed by modifying physical layout, post layout synthesis optimization is required.

14. Tapeout
    (a) When physical layout gate level netlist passes post layout GLS and post layout STA with the back annotated sdf, the design is ready for tapeout.
    (b) In tapeout phase, GDSII file is generated from layout and used for fabrication.

Synthesis

Synthesis is the process converting RTL code of a design to logic gates. Synthesis is dependent upon standard cell library, design constraints, and RTL code. The more cells in a standard cell library enable the synthesis tool to perform better optimization. However, if a standard cell library is too large with too many different flavors, it takes a lot of engineering resource to characterize the cells. Characterization of cells generates the .lib standard cell library which is used during synthesis.

Design constraints play an important role in synthesis. If a design is overconstraint, the synthesized circuit may be larger than necessary due to many parallel logic paths to meet the overconstraint timing requirements. If a design is underconstraint, the synthesized design may not meet timing performance and causes the design to fail. Therefore, realistic design constraints must be used during synthesis process to achieve optimal results.

As the synthesis process converts RTL code to logic gates, a badly coded RTL can lead the synthesis tool to synthesize redundant logic. This results in the synthesized circuit being larger than necessary. If the RTL contains code that is non synthesizable, the synthesis tool may synthesize garbage circuit thus causing functional failure in gate level netlist.

Standard Cell Library

A standard cell library is commonly in the format .lib (liberty format). It contains the timing information of every cell in the library. During synthesis, the synthesis tool refers to the standard cell library to synthesize the logic circuit from the RTL. Section "Digital Design Flow" shows some examples of different types of logic cells that are commonly present in the standard cell library. Each type of cell has different sizes which increase the flavors of the standard cell library. The more flavors it has, the better the synthesis optimization.

However a standard cell library that have too many flavors will require more engineering resource to characterize the standard cell library.

Figure 2.30 shows a diagram of the process flow for characterizing a standard cell library using Mentor Graphics' Kronos Characterizer tool. Each cell is designed using schematics and simulated using a spice simulator such as Mentor Graphics'

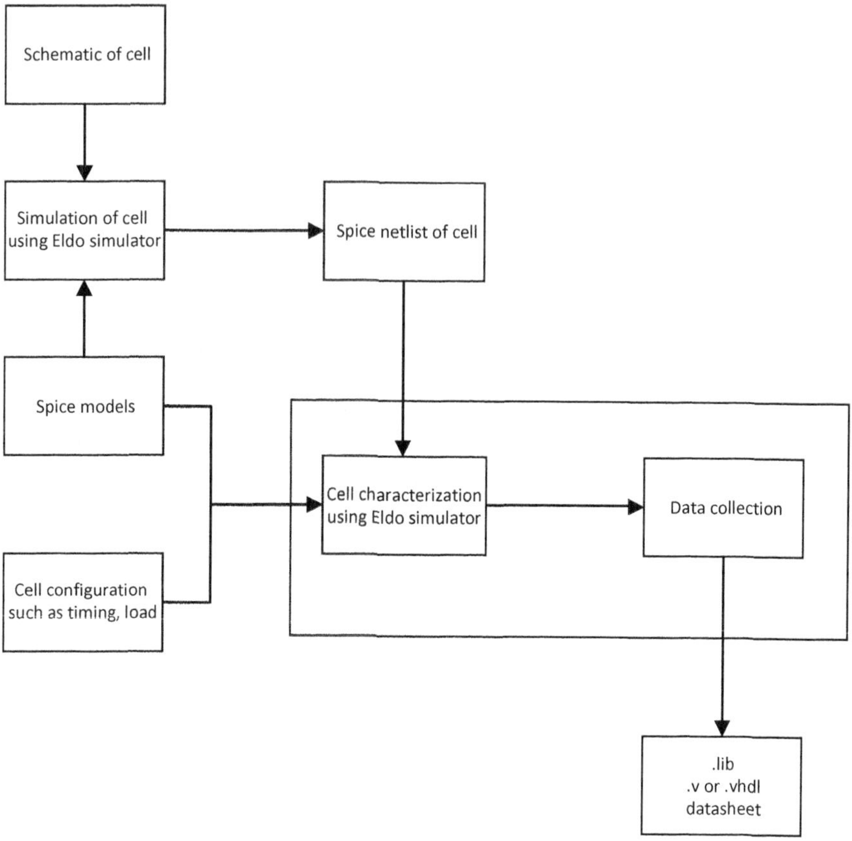

Fig. 2.30 Diagram showing process flow for characterization Of standard cell library using Mentor Graphics' Kronos Characterizer

Eldo. The spice netlist of the cell is then generated from the schematic and used as an input to the characterization tool. The spice model of the corresponding fabrication technology is also an input to the characterization tool. The cell configuration (timing, load) is another input to the characterization tool.

The characterization tool performs a series of simulation using different process corners from the spice model, different voltage, and different temperature. The series of simulation is measured, and the data collected and calculated to generate the corresponding liberty format (.lib), corresponding Verilog (or VHDL) library, and datasheet for the standard cell library.

The characterization tool generates a .lib and .v for three different scenarios, a min corner, a max corner, and a typical corner. The min corner refers to the fast corner which consists of the fast corner spice model, cold temp, and high voltage. For example, the cold temp is −40 °C, and high voltage is 3.6 V for a process that operates in the range of 2.7–3.6 V. The max corner refers to the slow corner which consists of the slow corner spice model, hot temp, and low voltage. An example is

100C for hot temp and 2.7 V for low voltage. The typical corner refers to the typical corner spice mode, typical temp, and typical voltage. An example is 27 °C temp and 3.3 V for typical voltage.

The characterization tool generates a _typ.lib, _min.lib, and _max.lib for liberty format and a _typ.v, _min.v, and _max.v for Verilog format.

The .lib file is used during synthesis, and the .v is used for gate level simulation.

Design Constraints

Design constraints are constraints that the designer uses to control the synthesis process. Design constraints are typically divided into:

- Input delay
- Output delay
- Path delay
- Clock requirement
- False path
- Multicycle path

It is important that the design constraints used during synthesis are realistic and not overconstraint or underconstraint. An overconstraint design will lead to synthesized circuit that is larger than it needs to be due to many parallel paths to meet timing. An underconstraint design will lead to synthesized circuit that is not able to meet timing requirements.

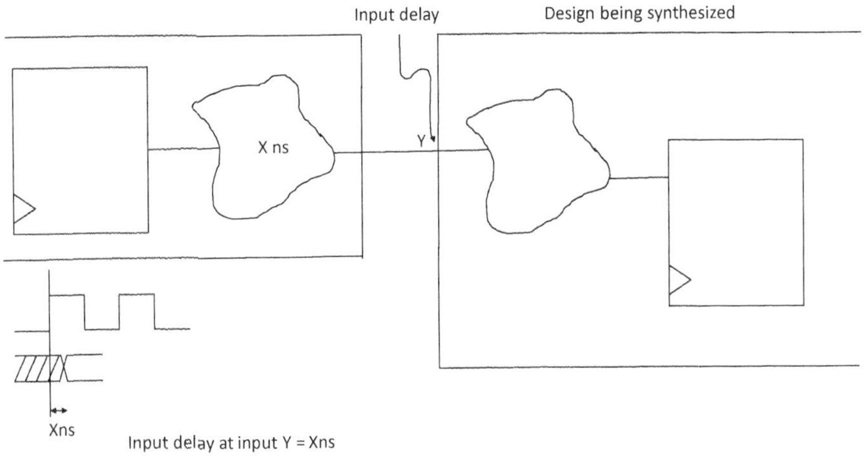

Fig. 2.31 Diagram showing input delay

Input Delay

Input to a design must be an output of another block or output of another chip. Therefore, an input delay refers to the amount of time after the active edge of previous clock in which the signal is valid at the corresponding output of the design.

Referring to Fig. 2.31, at the rising edge of clock, the signal goes through a combination of logic with a delay of xns before arriving at the input of the design being synthesized. Therefore, input Y of the design is only valid after *x* ns from rising edge of clock. The input delay to the design being synthesized is *x* ns.

Some design methodology or some design process may refer input delay as "in2reg" as "input to register." Some design tool such as Mentor Graphics' Leonardo Spectrum refers to input delay as arrival time.

Output Delay

Output from a design must be an input to another block or input to another chip. Therefore, an output delay refers to the amount of time before the active edge of next clock in which the signal must be valid at the corresponding input of the design.

Referring to Fig. 2.32, at the rising edge of clock, the signal Y goes through a combination logic in another block or another chip with a delay of x ns. Therefore, output Y of the design must be valid before x ns from rising edge of next clock.

Some design methodology or some design process may refer output delay as "reg2out," meaning "register to output." Some design tool such as Mentor Graphics' Leonardo Spectrum refers to output delay as required time.

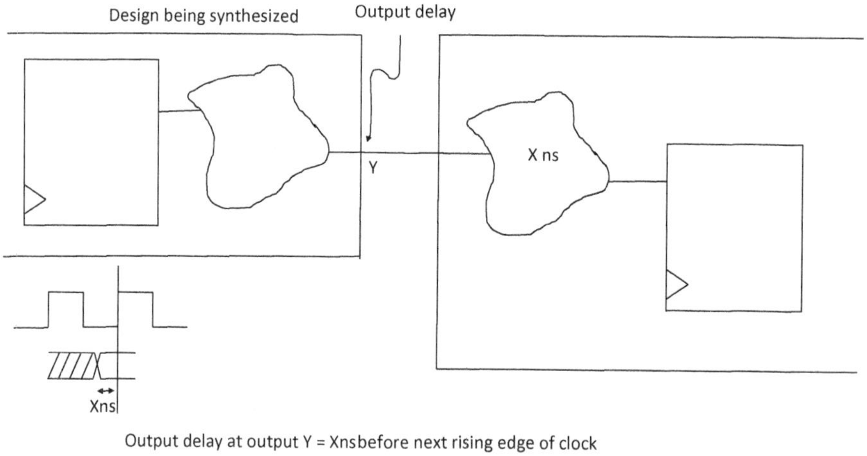

Fig. 2.32 Diagram showing output delay

Path Delay

During synthesis, the synthesis tool will try to optimize the logic based on the input delay, output delay, and clock frequency to obtain an optimized path for input to register and from register to output. For logic paths that do not involve registers and only combinational logic, path delay can be used to guide the synthesis tool to optimize the combinational paths. This can be achieved through setting of delay through a path. This is sometimes referred to as "in2out" for a combinational logic path (Fig. 2.33).

Clock Specification

Clock plays an important role in today's ASIC design. A clock specification can make a huge difference in the synthesis results of a design. If the clock is constrained too tightly and given an unrealistic clock frequency, the depth of the logic in the design will not be able to meet the requirement. If the clock is constrained too loosely, the logic will not be well optimized for timing.

Referring to Fig. 2.34, clock specification includes clock period (or clock frequency) and duty cycle. The shorter the clock period, the more difficult to optimize synthesis logic to meet timing requirements. A shorter clock cycle will result in more timing failures being setup time failures. The longer the clock period, the easier to optimize synthesized logic to meet setup time requirement but difficult to meet hold time requirements. Therefore, a realistic clock period must be used during synthesis. For example, if a design with complicated functionality with many levels of logic is expected, then based on an average delay per gate in the standard cell library, a maximum clock frequency can be estimated.

- Logic levels: Estimated at 30.
- Standard cell library: 0.35 μm with average delay per cell at 0.8 ns.
- Maximum estimated clock frequency for synthesis = 30 × 0.8 = 24 ns = ~40 MHz.

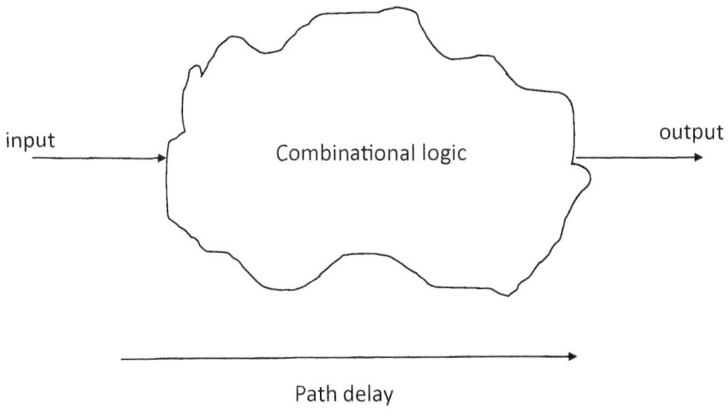

Fig. 2.33 Diagram showing path delay

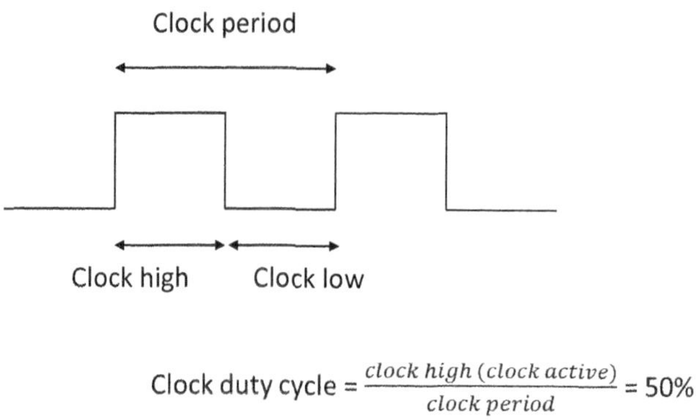

Fig. 2.34 Diagram showing clock specification

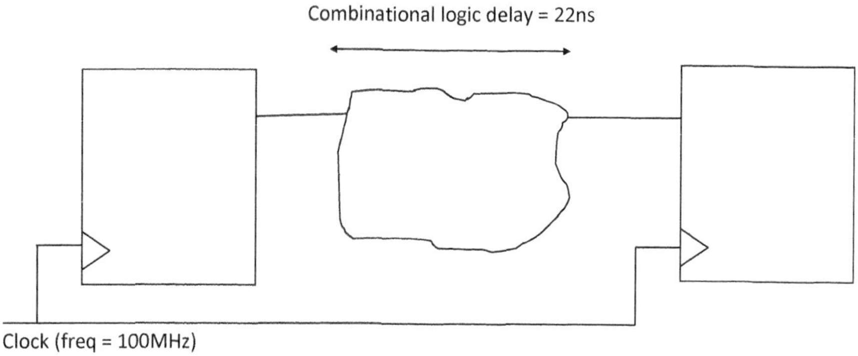

Fig. 2.35 Diagram showing multicycle path

Multicycle Path

A logic path with combinational logic that have its functional delay longer than one clock cycle is known as multicycle path. If the path is not specified as multicycle path, the synthesis tool will focus its effort to optimize the path but never able to optimize it to achieve the required timing. Therefore, to achieve optimal synthesis results, the designer must specify a multicycle path.

Figure 2.35 shows the diagram for a multicycle path. The combinational logic has a delay of 22 ns, while the clock has a clock period of 10 ns. The combinational logic will require three clocks to be valid at the output of the second flip-flop. Therefore, this path must be specified to the synthesis tool as a multiple cycle path with three clocks. Correctly configuring paths as multicycle paths can significantly improve the timing performance of a circuit. However, paths that are not multicycle in functionality cannot be specified as multicycle, lest those paths that fail timing will be reported as meet timing.

False Path

Synthesis process optimizes logic and functionality of a design to achieve best timing performance in order to meet specified timing. When a path fails timing, the synthesis tool will report the failing path for its setup failure or hold failure. The designer will have to determine how to fix the timing failure. However, there are paths that will always fail timing and do not require any fixes. This is typically paths that are asynchronous and can occur at any time (Fig. 2.36).

An example of a signal that is asynchronous and need not be considered during synthesis optimization is the reset signal. Asynchronous reset can occur at any time and does not have a timing reference to a clock. Therefore, during synthesis, it is declared as a false path.

Signals that are asynchronous in nature but not configured as false path will create unnecessary timing failure during synthesis and thereby misdirect the synthesis tool to optimize the unnecessary false path logic. However, a design cannot arbitrarily assign signals to be false path if they are not asynchronous. Wrongly assigning signals to be false path will result in real timing violations not optimized and not reported by the synthesis tool.

Synthesis Optimizations to Improve Timing

Chap. 1 shows some different standard cells that are commonly found in a standard cell library. Each functional standard cell will have different sizes catering for different drive strengths. Section "Design Constraints" has discussed about different design constraints such as input delay, output delay, clock frequency, false path, and

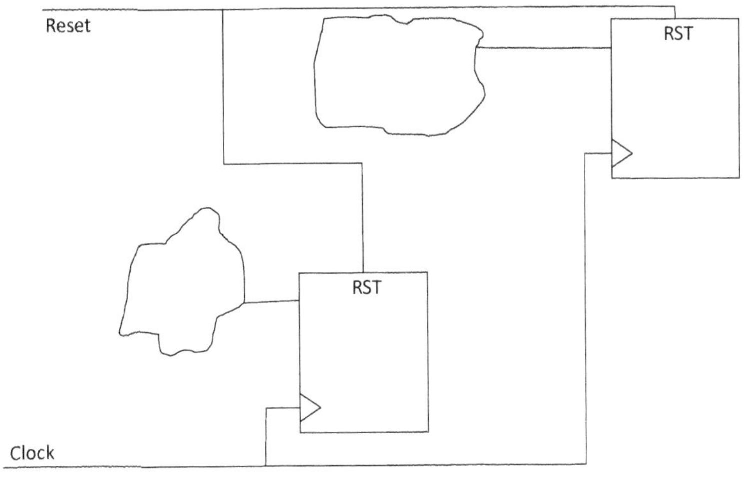

Fig. 2.36 Diagram showing false path

multicycle path. Both the standard cells in standard cell library and the design constraints play a major role during synthesis.

When a RTL code is being synthesized, its functionality as described in the RTL will convert to logic gates based on the available standard cells in the standard cell library. The more types of standard cells in a standard cell library will allow the synthesis tool to have better combinations of logic gates to achieve better circuit performance. However, a balance is needed as a standard cell library that is too large with too many standard cells is difficult to manage and requires a lot of resource to design the standard cells and characterize them into a standard cell library.

During synthesis, when converting the RTL code into logic gates, the synthesis tool uses the design constraint to optimize the synthesized logic to obtain better timing performance. Figure 2.37 shows a simple logic circuit with a late arriving signal D, utilizing priority encoding to optimize timing.

Signals A, B, and C arrive earlier compared to signal D. On the multiplex encoding, the worst path timing of 1.8 ns is due to the late arriving signal D at 1.2 ns, added with the 0.3 ns gate delay, resulting in a delay of 1.8 ns. If the design constraints specify the input delay of signals A, B, C, and D, the synthesis tool will know that signal D is a late arriving signal and optimize the circuit into priority encoding. In priority encoding, the late arriving signal is decoded last, with the early arriving signals A, B, and C being decoded earlier. This results in a longest path timing of 1.5 ns, saving 0.3 ns from the longest path of multiplex encoding.

Figure 2.38 shows a circuit which have a late arriving signal sel resulting in a longest path of 4.5 ns + 1 ns + 4 ns which adds up to 9.5 ns.

During synthesis, due to the design constraints showing late arriving signal of sel, the synthesis tool will optimize the circuit to duplicate the logic functionality after the multiplexer and duplicate it into two portions. This duplicated logic allows the early arriving signals of A and B to be decoded while waiting for the late arriving signal sel. When late arriving signal sel is valid, the multiplexer will multiplex the input to the output. As the logic functionality after the multiplexer have been duplicated and decoded for signal A and B, there is no logic functionality after the multiplexer. This results in a design that is larger due to the logic duplication resulting in same functional logic being used twice. However, this method is effective in improving timing, resulting in better performance.

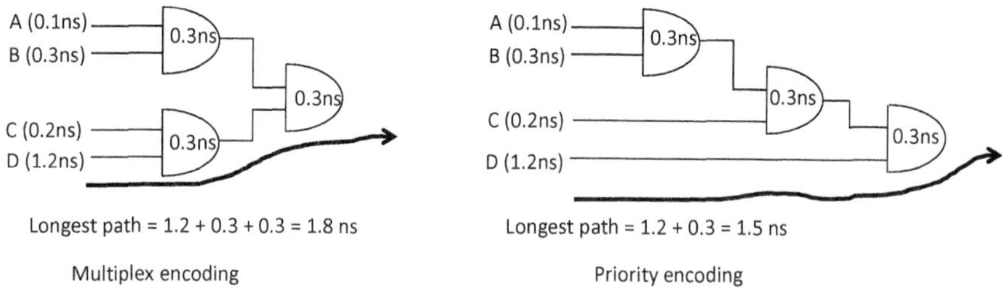

Fig. 2.37 Diagram showing priority encoding to improve timing

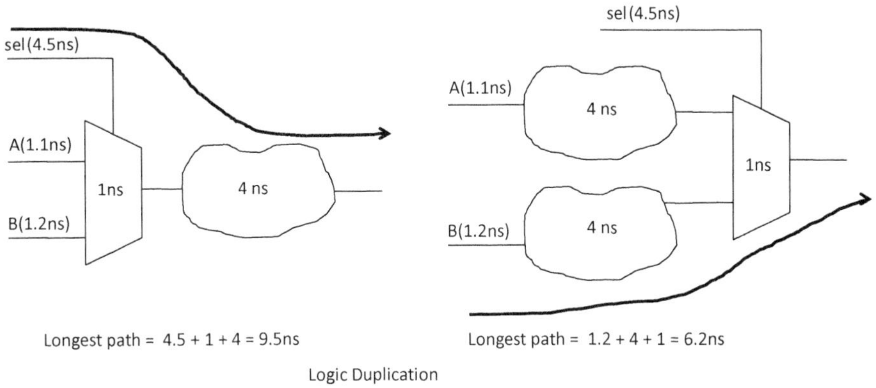

Fig. 2.38 Diagram showing logic duplication for slow arriving multiplexer select to improve timing

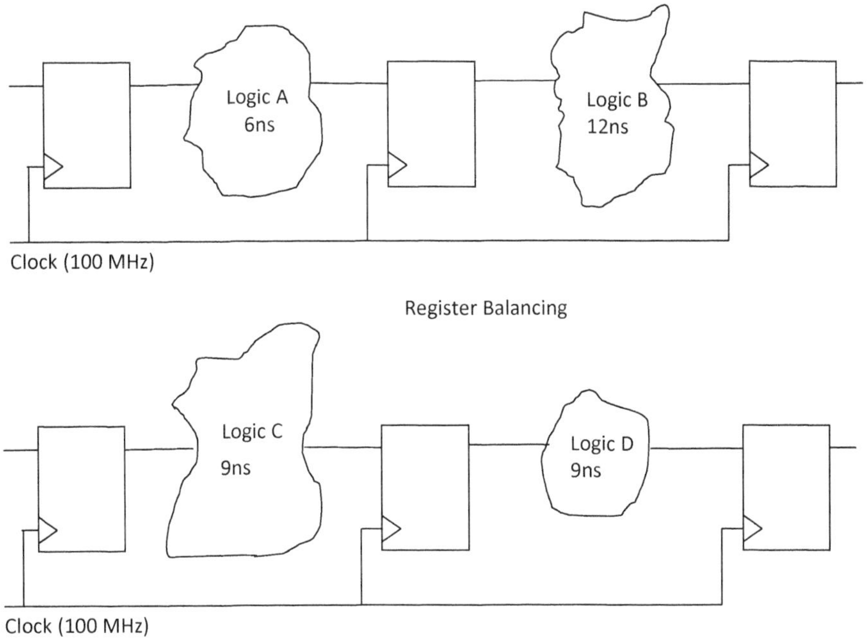

Fig. 2.39 Diagram showing register balancing to improve timing

Figure 2.39 shows a circuit with logic A on first stage of pipeline and logic B on second stage of the pipeline. Logic B has a longer path with a delay of 12 ns, while logic A has a shorter path with a delay of 6 ns. The clock frequency is 100 MHz which have a clock period of 10 ns.

Logic B cannot meet timing, and a setup failure occurs with a negative slack of 2 ns. In order to achieve timing, the synthesis tool can rebalance the pipeline and move some of the logic from logic B to logic A, forming logic C at the first stage increasing the delay to 9 ns while reducing the second stage delay to form logic D with 9 ns. The circuit is able to meeting timing with register balancing.

Figure 2.40 shows a circuit with logic A diverging to two paths of logic B and logic C.

Logic A can be duplicated to form two paths of logic A with logic B as one path, with the other path being logic A with logic C. Each path can be individually optimized to achieve better timing as opposed to logic A being shared and lack the flexibility for optimization. The disadvantage is that the circuit is now bigger with logic A being duplicated.

Importance of Clock in Backend

Design flow is typically separated into frontend and backend. Frontend is the process flow before physical layout and backend refers to process flow during physical layout to tapeout. During physical layout, the synthesized netlist is layout into physical layers such as poly, metal, active layers, and others to form the circuit and its interconnectivity. For digital design, this physical layout is commonly implemented using auto place & route.

Auto place & route places the cells in the netlist and connects the cells together using metal layers to form the complete circuit. Clock plays a very important role in physical layout.

Clock specification in synthesis typically involves clock period and clock duty cycle. These info are used by the synthesis tool to calculate timing and optimize the

Fig. 2.40 Diagram showing logic duplication and optimization to improve timing

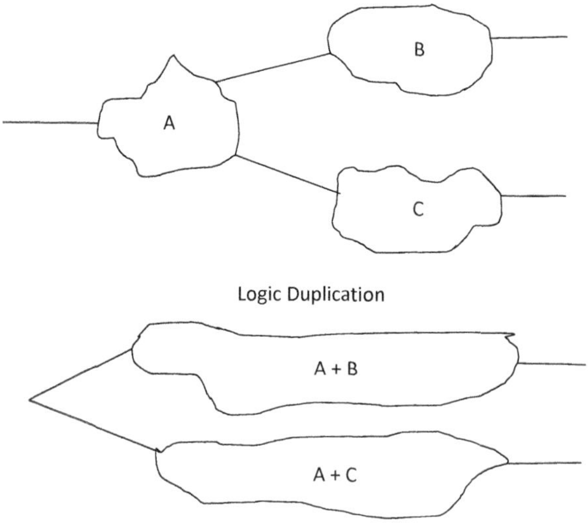

circuit. In layout, physical properties of the interconnectivity come into play and create a problem for the clock network.

Referring to Fig. 2.41, it shows clock signal routed to the clock port of two flip-flops at different physical locations. The routing for the clock to flip-flop B is longer than the routing for the clock to flip-flop A. This causes a problem as the clock will be seen active at flip-flop A earlier compared to flip-flop B. The maximum difference between the clock arriving at a flip-flop to the clock arriving at another flip-flop is called clock skew.

Clock skew can cause a circuit to have timing issues. If the clock travels a long clock tree to reach the destination flip-flop compared to the data path of the destination flip-flop, the data would have changed before it is latched in at the destination flop. This will cause a hold violation. If the clock travels a long clock tree to reach the source flip-flop compared to the destination flip-flop, the data at the destination flip-flop is not yet valid when the destination flip-flop triggers. This will cause a setup violation.

Floor Plan

Floor plan is the first step in physical layout. In floor planning, blocks of gates with similar functionality are placed together or near to each other. The objective is to achieve most optimal placements of cells and shortest possible routing. A good floor plan can significantly help the design achieve timing closure, while a bad floor plan can create long routing which increases delay and thereby causing timing violations.

Floorplan includes placement of IO pads in the chip configuration. It's important to have the IO pads in an optimal position on the chip to achieve optimal routing. Most auto place & route tools have built in features and functionality that can aid in the process of floor planning.

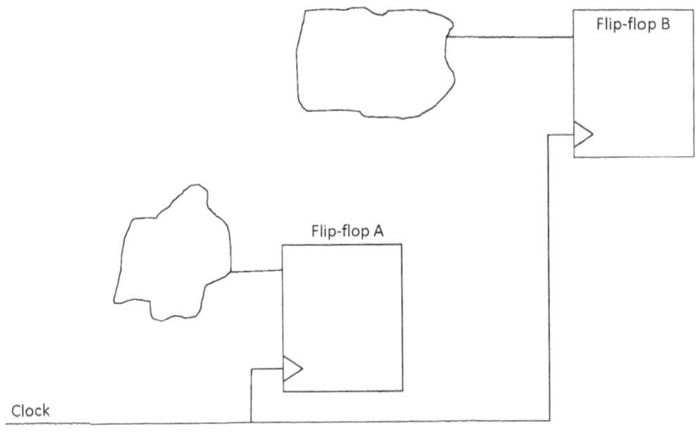

Fig. 2.41 Diagram showing clock routing to different flip-flops

For a large SOC or ASIC, if an IO pad that connects to a physical block is placed in a nonoptimal location, it will result in long routing lines which will increase the parasitics and slowing the signal, as shown in Fig. 2.42. As the long line takes up more space in fullchip to route from the nonoptimal IO pad location to its interconnect physical block, it will reduce the available space for other signal routings. This in turn increases the inefficiency of other routing signals and further increases the delay of those signals.

As there is less available space, thus the area utilization is also impacted and will result in larger than necessary area which again increases the parasitics which in turn increases the delay of other signals. A badly placed IO not only affects the signal which it routes to the corresponding logic block but it affects other fullchip signal routings as well. For a fullchip with several badly places IO pads, it can lead to signal congestion on fullchip.

In floor plan, you may have two situations which are commonly referred to as pad limited and core limited. Figure 2.43 shows a pad limited design and a core limited design.

In pad limited design, the number of IO pads requires a much larger space compared to the core design. Area is a nonissue as there is larger area than is required by the core resulting in underutilization of space. For the design which have large amount of pinouts resulting in pad limited, if area is a constraint, designers can opt for staggered bond as shown in Fig. 2.44, which can reduce the fullchip area size, but with added wire bonding complexity.

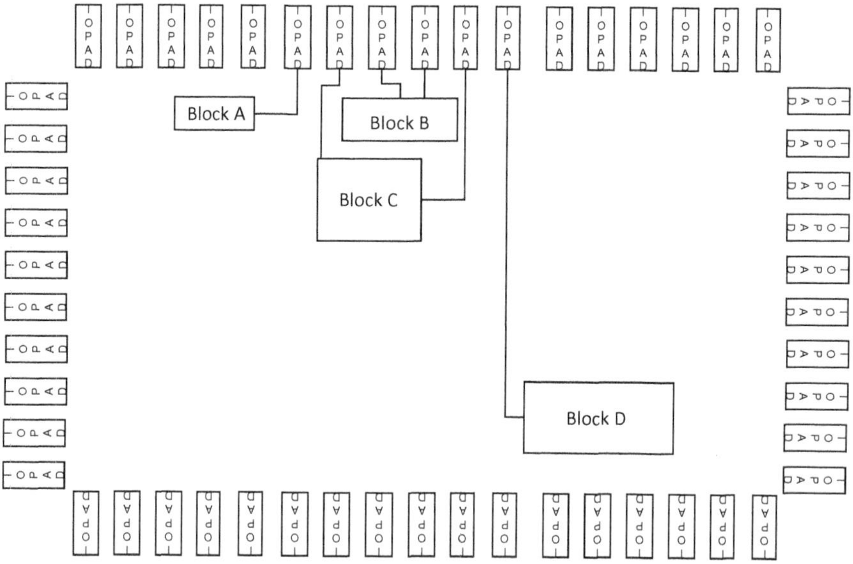

Fig. 2.42 Diagram showing long fullchip routing line due to a badly placed IO pad

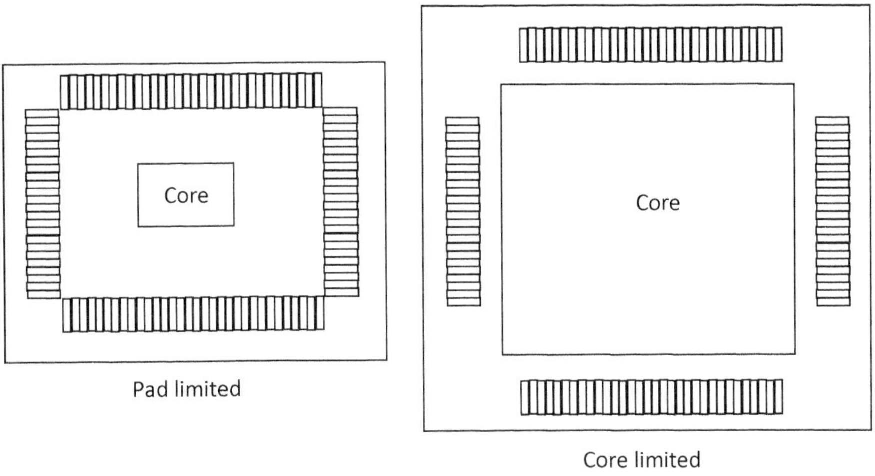

Fig. 2.43 Diagram showing pad limited and core limited design

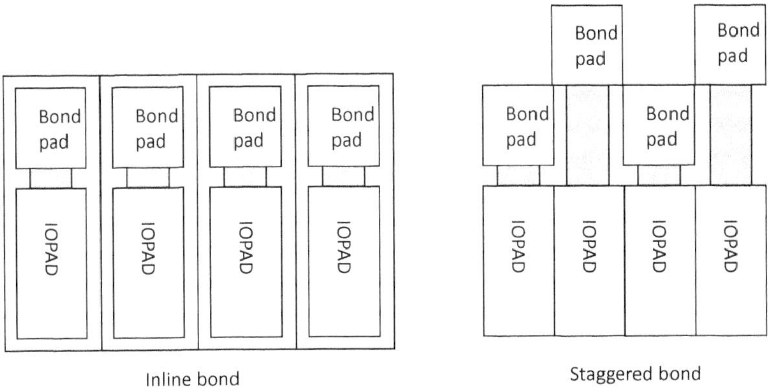

Fig. 2.44 Diagram showing inline bond and staggered bond

For core limited design, the core area is larger than that required by the pads; thus, the core area determines the size of fullchip. For such cases, optimal core design, core synthesis, and core routing play a larger role in determining die size.

Apart from the design module interconnects and the IO pad location, another aspect that plays a role in floor planning is the power mesh. The power mesh can be considered a pre-route as power mesh is routed with top metal layers prior to placement and routing of standard cells. Power mesh is important to ensure IR drop is kept to a minimal, and electromigration is avoided.

For a large SOC or ASIC design, when current travels a VDD line through a metal layer to provide VDD to different parts of a chip, the parasitics from the metal layer can cause a voltage drop or commonly referred to as IR drop. When there is a significant IR drop, the chip slows down thus not able to meet its speed requirement

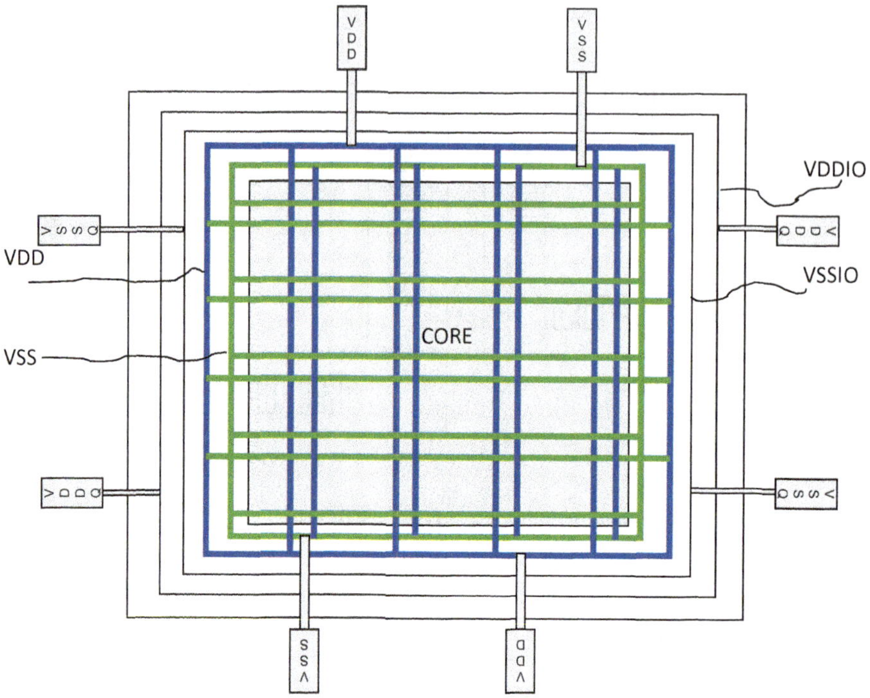

Fig. 2.45 Diagram showing VDDQ, VSSQ, VDD, and VSS

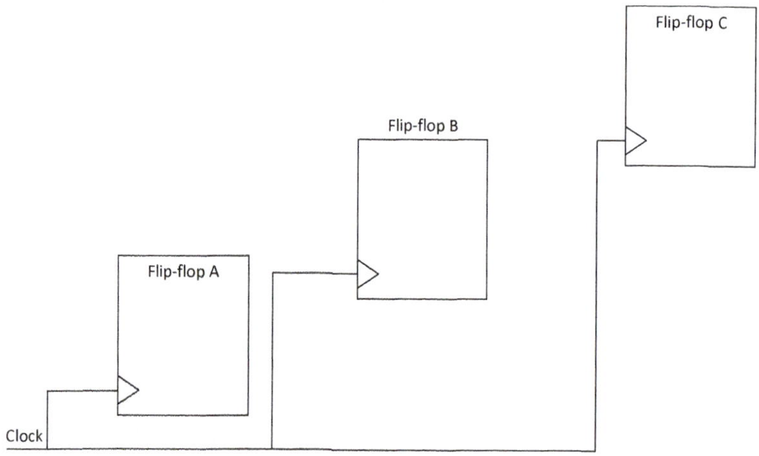

Fig. 2.46 Diagram showing clock tree before balance

and timing failures resulting in functionality issues. Apart from IR drop, when large current travels through a metal layer, it may result in the electrons from the high current colliding with the metal atoms causing the atoms to move. In severe cases,

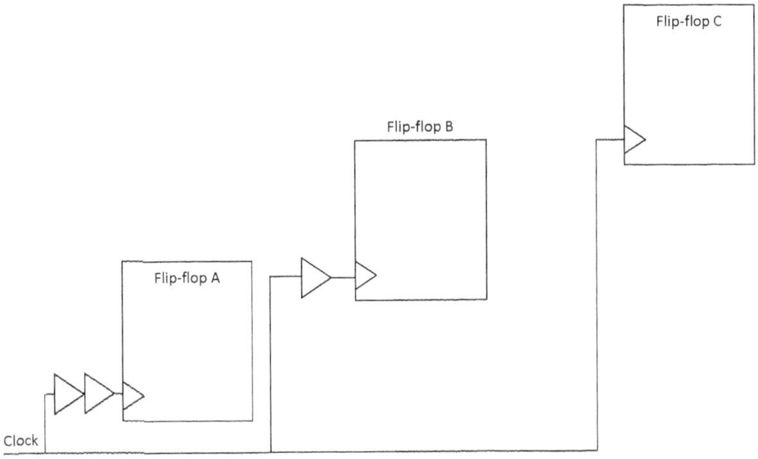

Fig. 2.47 Diagram showing clock tree after balancing

so much metal atoms have moved that the metal becomes an open or short resulting in chip failure. This is referred to as electromigration.

To avoid IR drop issues, designers commonly assign power rings with straps and separating the core power/ground from the IO pad power/ground, as shown in Fig. 2.45. Commonly as many IO pads as possible that are unused are catered for power and ground to ensure there is adequate IO pads at different locations to minimize the voltage drop. Horizontal and vertical power straps and ground straps are run across the core to minimize the impact of IR drop. Figure 2.45 shows the top two metal layers are run across with one top metal layer for vertical and another top metal layer for horizontal. Both these two top metals run across the core to strap the power and ground to minimize the IR drop.

The IR drop and electromigration requirements are dependent on the fabrication process technology. In planning the power rails, it is important to calculate the required metal width to avoid IR drop and electromigration. Once the power rails have been created, it is a good practice to run power analysis tools to verify the power rails meet the requirements of IR drop and electromigration.

Clock Tree Synthesis

After floor plan and the design are placed, clock tree synthesis is implemented. In this step, the clock network is routed with highest priority in order to achieve a balanced tree network.

Figure 2.46 shows a design with three flip-flops before the clock tree is balanced. Figure 2.47 shows the same design after the clock tree is balanced. Flip-flop A have its clock signal inserted with two clock buffers to delay the clock signal from arriving at the clock pin of flip-flop A. Flip-flop B have its clock signal inserted with one clock buffer, and flip-flop C does not have any clock buffer inserted into its clock signal. This balancing tries to balance the clock to arrive at the clock pin of all three flip-flops together in order to minimize clock skew.

It is impossible to achieve zero clock skew. Clock tree synthesis will insert series of different sizes of clock buffers to achieve an optimal clock tree with minimal clock skew.

Chapter 3
Multiple Clock Design

ASIC and SoC have multiple functionality and multiple features design onto the same chip. The main advantage of an ASIC or SoC is an efficient design with lower cost and better performance. Cost advantage is achieved by having fewer chips in a system and thus reducing the cost of multiple chip packaging and smaller PCB footprint on system level. Performance advantage is achieved by having short metal routings on chip level instead of copper interconnects between multiple chips on PCB at system level.

However, a major disadvantage from a design perspective is the existence of multiple clock domains on the chip. For example, Fig. 3.1 shows a SoC chip with multiple functionalities. Each functionality may have its own clock, therefore creating multiple clock domains on the SoC chip. Interaction between logic of one clock domain to logic of another clock domain can cause signals to be corrupted if the clock domain is asynchronous. This signal corruption is referred to as clock domain crossing or CDC.

CDC only occurs if the signal crosses from an asynchronous clock domain to another asynchronous clock domain. CDC does not occur if the signal crosses between synchronous clock domain. Synchronous clock domain refers to transmitting clock domain, and receiving clock domain is from the same source. Asynchronous clock domain refers to transmitting clock domain, and receiving clock domain is asynchronous with no relation between them. For asynchronous clock, there is a possibility that the transmitting clock may trigger within the setup and hold time domain of the receiving clock. When this occurs, there is a metastable condition in the receiving flip-flop.

Referring to Fig. 3.2, the signal crosses from clock1 domain to clock2 domain. Both clocks are asynchronous and can occur anytime. There is a possibility that clock1 rising edge will be within the setup and hold time region of clock2. When this occurs, flip-flop B will not be able to capture the signal due to metastability. Signal corruption due to CDC cannot be avoided and must be addressed in hardware RTL design.

W. F. Lee, *Learning from VLSI Design Experience*,
https://doi.org/10.1007/978-3-030-03238-8_3

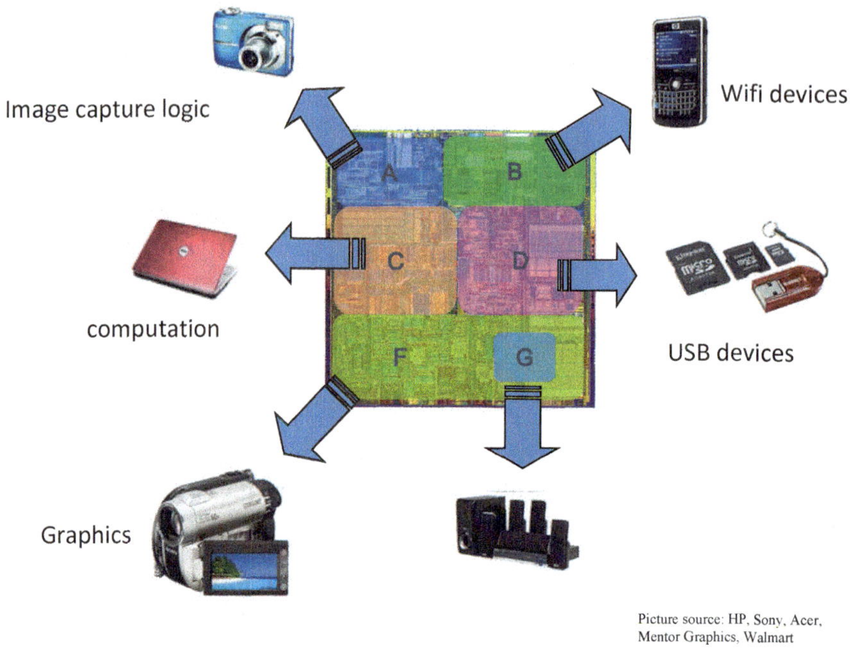

Fig. 3.1 Diagram showing SoC with multiple functionality

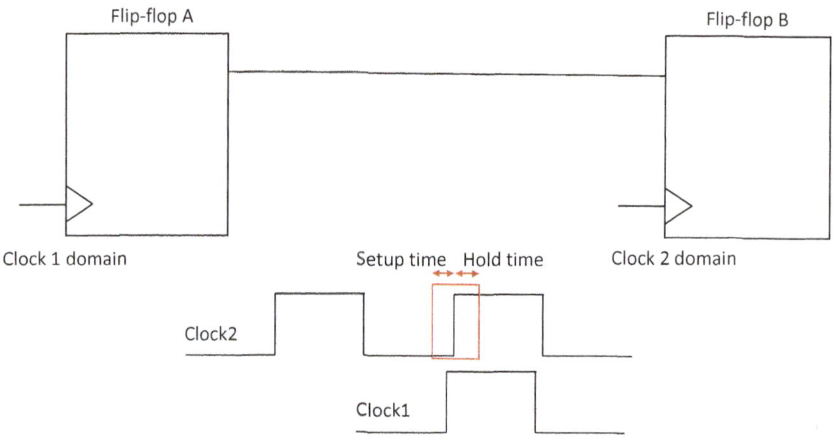

Fig. 3.2 Diagram showing signal crossing asynchronous clock domain

Mean Time Between Failure

CDC issues cannot be avoided when a signal crosses from one asynchronous clock domain to another. There is a possibility on clock1 rising edge triggers within the setup and hold time region of clock2, thus causing a metastability. There are several equations available in academic literature to determine MTBF (mean time between failure) for CDC failure, a commonly used equation:

$$\text{MTBF} = \frac{e^{\frac{t}{\tau}}}{f_{\text{clk1}} f_{\text{clk2}} K}$$

t represents the resolution time or time taken for mestastability to resolve, τ represents the time constant, and K is the metastability window. Referring to Fig. 3.2, if the transmitting clock triggers within the metastability window of receiving clock, a failure occurs. The larger the metastability window, the smaller the mean time between failure. The setup time requirement and hold time requirement of the receiving flip-flop determine the metastability window. A fast flip-flop will have a smaller metastability window thus allowing for a larger mean time between failure.

For a given technology, these variables for MTBF are fixed based upon the design of the flip-flop standard cell and the technology process involved. For example, a flip-flop standard cell on a 0.18 μm technology might have t = 2.0 ns, τ = 0.25 ns, and K = 10 as (1 × 10^{-18} s), assuming f_{clk1} = 10 MHz and f_{clk2} = 100 MHz

$$\begin{aligned} \text{MTBF} &= \frac{e^{\frac{2.0}{0.25}}}{1 \times 100 \times 10^{6} \times 10 \times 10^{6} \times 10 \times 10^{-18}} \\ &= \frac{2980.957987}{10 \times 10^{-3}} \\ &=\sim 298{,}096\,\text{s} \\ &=\sim 82.8\,\text{h} \\ &=\sim 3.45\,days \end{aligned}$$

This means a potential failure due to metastability will occur in ~3.45 days if the design is used for a period longer than that. Metastability failures are intermittent, thus making debugging difficult as they are difficult to reproduce.

Metastability due to CDC cannot be avoided and is part of the nature of digital design. However, it can be work-around by increasing the MTBF to a value large enough such that it will never occur physically. Figure 3.3 shows an additional flip-flop C added to the clock2 domain.

The effective MTBF now increases due to flip-flop C. In physical layout, flip-flop C is placed next to flip-flop B to minimize the routing delay between flip-flop B and flip-flop C. There cannot be any logic between flip-flop B and flip-flop C. The presence of logic will increase the delay between flip-flop B and flip-flop C, thereby reducing the effective MTBF.

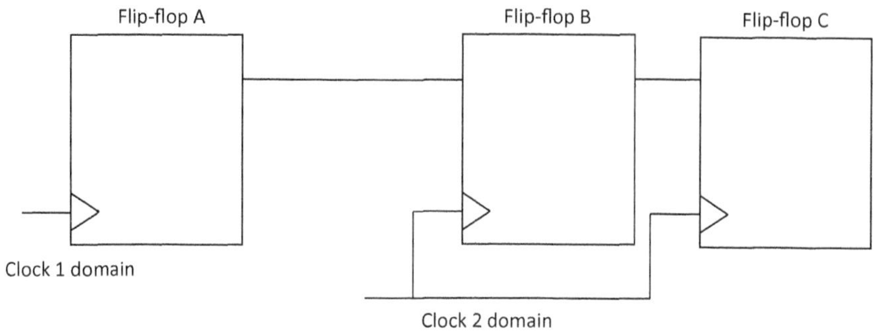

Fig. 3.3 Diagram showing flip-flop C added to clock2 domain to increase MTBF

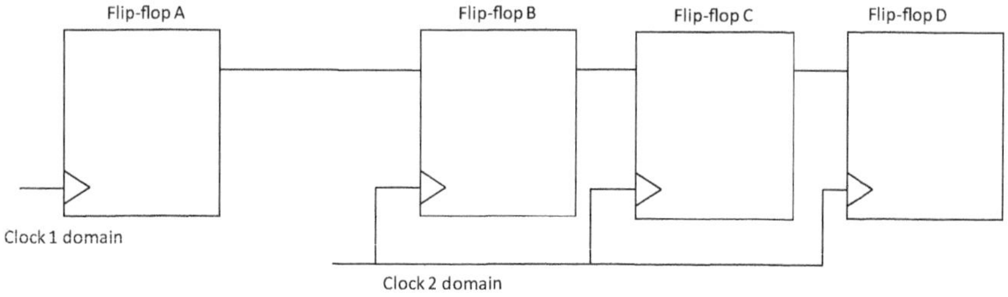

Fig. 3.4 Diagram showing three-flip-flop synchronizer to increase MTBF

$$
\begin{aligned}
\text{MTBF} \quad &= \text{MTBF}_{\text{flip flopB}} \times \text{MTBF}_{\text{flip flopC}} \\
&=\sim \left(\text{MTBF}_{\text{flip flopB}}\right)^2 \\
&=\sim (82.8\,\text{h})^2 \\
&=\sim 6{,}855.84\,\text{h} \\
&=\sim 285.66\,\text{days}
\end{aligned}
$$

If this is not adequate, another flip-flop can be added to clock2 domain, forming a three-flip-flop synchronizer as shown in Fig. 3.4.

$$
\begin{aligned}
\text{MTBF} \quad &= \text{MTBF}_{\text{flip flopB}} \times \text{MTBF}_{\text{flip flopC}} \times \text{MTBF}_{\text{flip flopD}} \\
&=\sim \left(\text{MTBF}_{\text{flip flopB}}\right)^3 \\
&=\sim (82.8\,\text{h})^2 \\
&=\sim 567{,}663.55\,\text{h} \\
&=\sim 23{,}652.65\,\text{days} \\
&=\sim 64.8\,\text{years}
\end{aligned}
$$

By having three flip-flops, the MTBF is increased to 64.8 years for the digital circuit. MTBF is dependent upon the process technology that governs the characteristics of the flip-flop and the frequency of clock1 and clock2. Typically, for most application, a two-flip-flop synchronizer is adequate to increase the MTBF to a reasonable value. However, for some circuits, a three-flip-flop synchronizer may be required. It is recommended that designers always make a calculation on the MTBF to determine how many flip-flops are required for the synchronizer to ensure the MTBF is large enough for the circuit's application.

Synchronizer

When a signal crosses from one asynchronous clock domain to another, a metastability may occur, and the probability of failure can be determined by the MTBF (mean time between failure) equation.

To increase the MTBF to ensure the metastability failure will not occur within a practical timeframe of usage of the design, synchronizer is added into the circuit at the receiving end, as shown in Fig. 3.3.

When adding synchronizer to a design with clock domain crossing, two conditions need to be considered:

a. Receiving clock faster than transmitting clock
b. Transmitting clock faster than receiving clock

Receiving Clock Faster than Transmitting Clock

When a signal crosses from an asynchronous clock domain to another asynchronous clock domain, adding a two-flip-flop synchronizer changes the functionality as the signal at receiving clock is delayed by two clocks. Figure 3.5 shows a signal

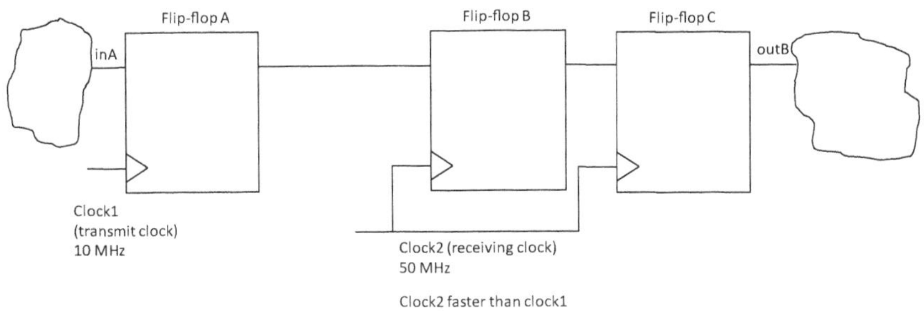

Fig. 3.5 Diagram showing two-flip-flop synchronizer from a slower transmitting clock to a faster receiving clock

crossing from asynchronous clock1 domain to clock2 domain with a two-flip-flop synchronizer.

Example 3.1 shows the Verilog code of a module with a two-flip-flop synchronizer, while Example 3.2 shows a Verilog testbench that is used for simulation. Figure 3.6 shows the simulation results without and with the two-flip-flop synchronizer. OutB signal is delayed by two clocks with the two-flip-flop synchronizer. Although synchronizer increases MTBF and is required for CDC, the designer needs to be aware that the corresponding signal is delayed by two clocks once a two-flip-flop synchronizer is added into the design.

Example 3.1 Verilog code showing implementation of two flip-flop synchronizer

```
module cdc1 (inA, outB, clock1, clock2, reset);
input inA;
input clock1; // 10 MHz
input clock2; // 50 MHz
input reset;
output outB;
reg outB;
reg int_inA;

reg int_inAq1, int_inAq2;


always @ (posedge clock1 or posedge reset)
begin
       if (reset)
```

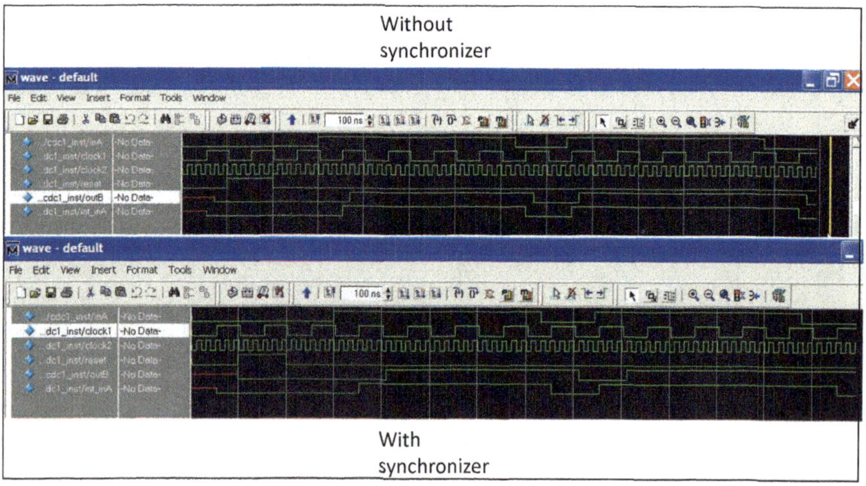

Fig. 3.6 Diagram showing simulation results of cdc1_tb

```
                int_inA <= 0;
        else
                int_inA <= inA;
end

always @ (posedge clock2 or posedge reset)
begin
        if (reset)
                begin
                        int_inAq1 <= 0;
                        int_inAq2 <= 0;
                        outB <= 0;
                end
        else
                begin
                        int_inAq1 <= int_inA;
                        int_inAq2 <= int_inAq1;
                        outB <= int_inAq2;
                end
end
endmodule
```

Example 3.2 Verilog testbench to simulate module cdc1

```
module cdc1_tb ();

reg inA;
reg clock1;
reg clock2;
reg reset;
wire outB;
reg [3:0] i,j,k,l;

initial
begin
        clock1 = 0;
        j = $random;
        #j;
        forever #50 clock1 = ~clock1; // 10 MHz
end

initial
begin
        clock2 = 0;
        k = $random;
        #k;
```

```
        forever #10 clock2 = ~clock2; // 50 MHz
end
initial
begin
        reset = 0;
        inA = 0;
        #100;
        reset = 1;
        #100;
        reset = 0;
        #100;
        for (i=0; i<10; i=i+1)
        begin
                inA = $random;
                l = $random;
                #(l+100);
        end
        $stop;
end

cdc1 cdc1_inst (.inA(inA), .outB(outB), .clock1(clock1), .
clock2(clock2), .reset(reset));

endmodule
```

Transmitting Clock Faster than Receiving Clock

When a signal crosses from an asynchronous clock domain to another, adding a synchronizer can increase the MTBF of the metastability failure, but if the transmitting clock is faster than the receiving clock, a condition of lost data will occur. This happens due to the transmitting data changes faster than the receiving synchronizer is able to latch in the data. When the receiving synchronizer is latching in a data, the transmitting clock changes and drives new data. Since the receiving synchronizer is

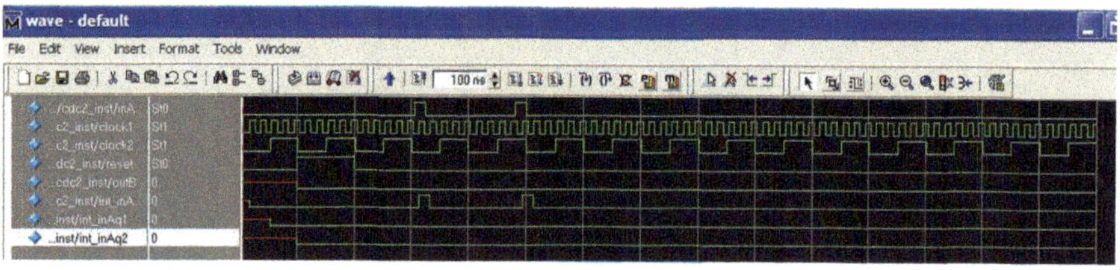

Fig. 3.7 Diagram showing OutB does not capture the pulses at inA

running on a slower clock, it will miss some of the data. Therefore, adding a synchronizer alone is not recommended when crossing from a faster transmitting clock to a slower receiving clock.

Example 3.3 shows the Verilog testbench to simulate the Verilog code of Example 3.1, with clock1 at 50 MHz and clock2 at 10 MHz, thus creating a condition of transmitting clock faster than receiving clock. Figure 3.7 shows the simulation results, whereby the pulses at inA are not captured correctly at outB.

Example 3.3 Verilog testbench to simulate transmit clock faster than receive clock

```
module cdc2_tb ();
reg inA;
reg clock1;
reg clock2;
reg reset;
wire outB;
reg [3:0] i,j,k,l;

initial
begin
        clock1 = 0;
        j = $random;
        #j;
        forever #10 clock1 = ~clock1; // 50 MHz
end

initial
begin
        clock2 = 0;
        k = $random;
        #k;
        forever #50 clock2 = ~clock2; // 10 MHz
end
initial
begin
        reset = 0;
        inA = 0;
        #100;
        reset = 1;
        #100;
        reset = 0;
        #100;
        #6;
        inA = 1;
        #20;
```

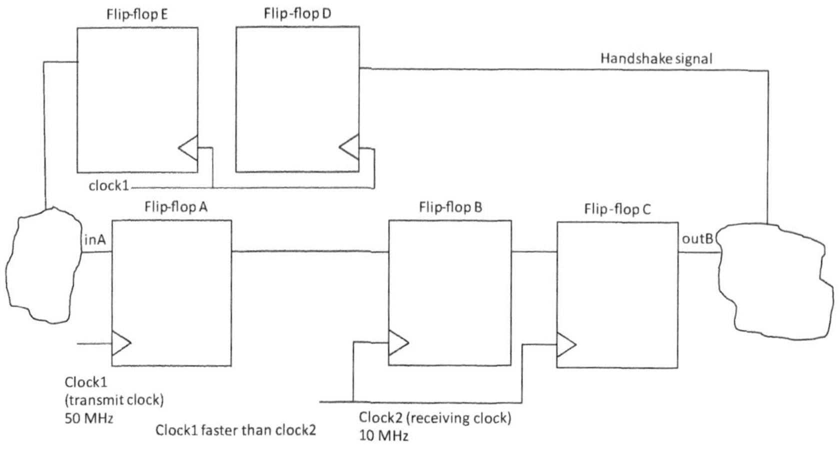

Fig. 3.8 Diagram showing handshake signal for CDC from fast transmit clock to slower receive clock

```
        inA = 0;
        #155;
        inA = 1;
        #20;
        inA = 0;
        #1000;
        $stop;
end

cdc1   cdc1_inst   (.inA(inA),   .outB(outB),   .clock1(clock1),   .
clock2(clock2), .reset(reset));


endmodule
```

A solution to this would be to create a handshake signal that goes from the receiving side to the transmitting side which can signal to the transmitting side to hold its data and not switch to a new set of data until the receiving side is ready. Such a handshake signal will cross from asynchronous clock2 domain to clock1 domain and will therefore need a synchronizer as shown in Fig. 3.8.

A commonly used solution is to create a toggle flag within the receiving domain. The toggle flag signal will toggle whenever inA is a logic one. The toggle flag will go through a synchronizer and xor logic to generate outB. Example 3.4 shows the Verilog code for such a design.

Example 3.4 Verilog code showing usage of toggle flag to generate outB for transmit clock faster than receive clock

```
module cdc2 (inA, outB, clock1, clock2, reset);
input inA;
input clock1; // 50 MHz
input clock2; // 10 MHz
input reset;
output outB;
reg outB;
reg int_inA;
reg flag_int_inA;
reg outBq1;
reg outBq2;
reg outBq3;

always @ (posedge clock1 or posedge reset)
begin
       if (reset)
              int_inA <= 0;
       else
              int_inA <= inA;
end

always @ (posedge clock1 or posedge reset)
begin
       if (reset)
              flag_int_inA <= 0;
       else if (int_inA)
              flag_int_inA <= ~flag_int_inA;
end
always @ (posedge clock2 or posedge reset)
begin
        if (reset)
```

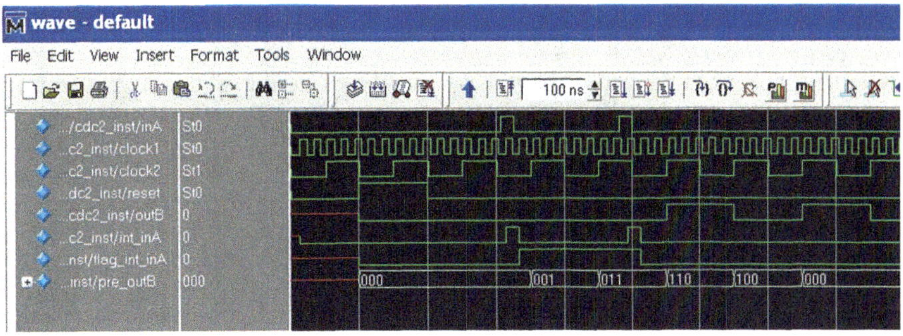

Fig. 3.9 Diagram showing simulation results of using toggle flag to capture inA when moving from faster transmit clock to slower receive clock

```
        begin
                outBq1 <= 0;
                outBq2 <= 0;
                outBq3 <= 0;
                outB <= 0;
        end
        else
        begin
                outBq1 <= flag_int_inA;
                outBq2 <= outBq1;
                outBq3 <= outBq2;
                outB <= outBq3 ^ outBq2;
        end
end
endmodule
```

Using the Verilog testbench of Example 3.3 with instantiation of design under test being module cdc2 of Example 3.4, Fig. 3.9 shows the simulation results.

Referring to Fig. 3.9, the internal signal flag_int_inA will toggle when int_inA is a logic 1. On the first pulse of inA, on rising edge of clock1, int_inA goes high, the next rising edge of clock1 will see flag_int_inA toggle from low to high. On the second pulse of inA, next rising edge of clock1, int_inA goes high, with the next rising edge of clock1 seeing flag_int_inA toggle from high to low. This creates a logic 1 pulse between the two pulses of int_inA. This pulse of flag_int_inA then goes through a three-flip-flop synchronizer to generate outBq1, outBq2, and outBq3, each being one clock2 cycle later than the next. OutB is then generated from outBq2 and outBq3 in order to achieve the two pulses of inA but synchronized to clock2 domain.

OutB is generated from xor logic of outBq2 and outBq3, but not outBq1 and outBq2. Both would yield the same functionality except the former will be one

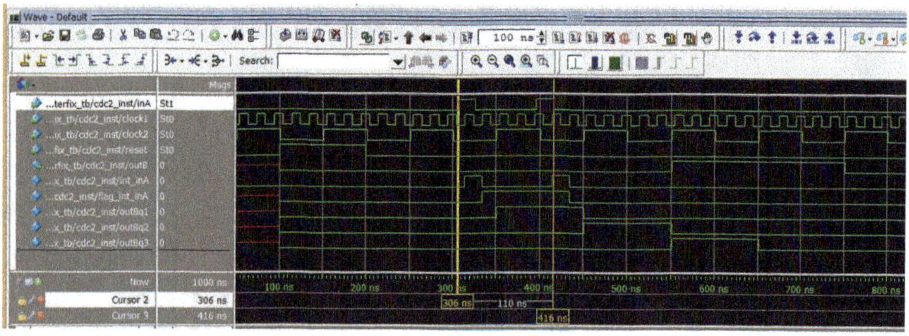

Fig. 3.10 Diagram showing simulation results of using shorter time between inA pulse when transmit clock faster than receive clock

clock cycle later. But outBq1 is only one-flip-flop synchronizer; thus, its MTBF will not be as good as xor logic of outBq2 and outBq3.

Referring to Fig. 3.9, the two pulses of int_inA are separated by three edges of clock2 (rising edge clock2, falling edge clock2, and rising edge clock2). If the two pulses of int_inA are brought closer, it will reach a point where the two pulses of int_inA will not be captured at outB. Verilog code shown in Example 3.5 shows the testbench pumping in stimulus with a shorter time between inA pulses. Figure 3.10 shows the simulation results whereby outB is captured as a single pulse instead of two pulses.

Example 3.5 Testbench to simulate with shorter time between inA pulse when transmit clock is faster than receive clock

```
module cdc2_tb ();
reg inA;
reg clock1;
reg clock2;
reg reset;
wire outB;
reg [3:0] i,j,k,l;

initial
begin
        clock1 = 0;
        j = $random;
        #j;
        forever #10 clock1 = ~clock1; // 50 MHz
end

initial
begin
        clock2 = 0;
        k = $random;
        #k;
        forever #50 clock2 = ~clock2; // 10 MHz
end

initial
begin
        reset = 0;
        inA = 0;
        #100;
        reset = 1;
        #100;
        reset = 0;
```

```
        #100;
        #6;
        inA = 1;
        #20;
        inA = 0;
        #70; // #155;
        inA = 1;
        #20;
        inA = 0;
        #1000;
        $stop;
end

cdc2  cdc2_inst  (.inA(inA),  .outB(outB),  .clock1(clock1),  .
clock2(clock2), .reset(reset));


endmodule
```

Referring to Fig. 3.10, the synchronizer is not able to capture the two pulses of inA as there is too little time between the two pulses of inA. A good reference to ensure successful capture by the synchronizer is a minimum of three clock2 edges between the two pulses of inA.

If the design requires a handshake signal to go from clock2 domain back to clock1 domain, the design shown in Example 3.4 can be further expanded. The handshake signal going to clock1 domain must also go through a synchronizer in clock1.

The Verilog code shown in Example 3.6 shows the outBq2 signal from the receive domain goes to transmit domain and goes through a synchronizer clocked by clock1 to generate the signal RX_busyq2. This signal is then xor with the toggle flag signal flag_int_inA to generate the RX_busy signal.

Example 3.6 Verilog code showing handshake signal from receive domain to transmit domain

```
module cdc3 (inA, outB, clock1, clock2, reset, RX_busy);
input inA;
input clock1; // 50 MHz
input clock2; // 10 MHz
input reset;
output outB;
output RX_busy;
reg outB;
reg int_inA;
reg flag_int_inA;
reg outBq1;
```

```
reg outBq2;
reg outBq3;
reg RX_busy;
reg RX_busyq1;
reg RX_busyq2;

always @ (posedge clock1 or posedge reset)
begin
        if (reset)
                begin
                        int_inA <= 0;
                        RX_busyq1 <= 0;
                        RX_busyq2 <= 0;
                        RX_busy <= 0;
                end
        else
                begin
                        int_inA <= inA;
                        RX_busyq1 <= outBq2;
                        RX_busyq2 <= RX_busyq1;
                        RX_busy <= flag_int_inA ^ RX_busyq2;
                end
end

always @ (posedge clock1 or posedge reset)
begin
        if (reset)
                flag_int_inA <= 0;
```

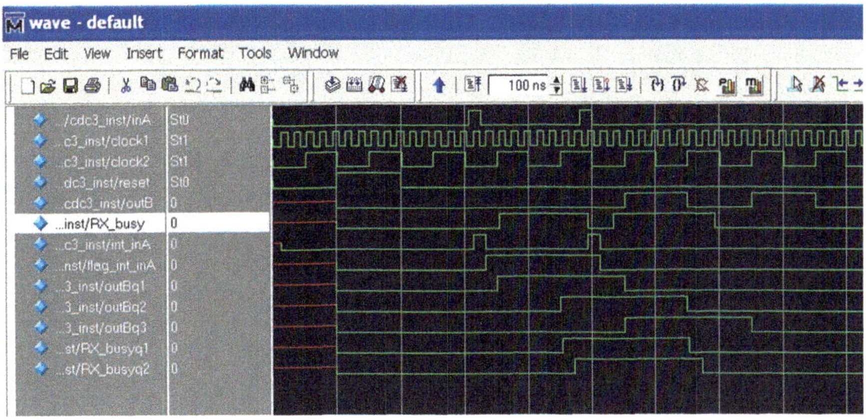

Fig. 3.11 Diagram showing simulation results of RX_busy when transmit clock faster than receive clock

```
        else if (int_inA & ~RX_busy)
                flag_int_inA <= ~flag_int_inA;
end

always @ (posedge clock2 or posedge reset)
begin
        if (reset)
                begin
                        outBq1 <= 0;
                        outBq2 <= 0;
                        outBq3 <= 0;
                        outB <= 0;
                end
        else
                begin
                        outBq1 <= flag_int_inA;
                        outBq2 <= outBq1;
                        outBq3 <= outBq2;
                        outB <= outBq3 ^ outBq2;
                end
end

endmodule
```

Example 3.7 shows the Verilog testbench for simulating the cdc3 module, with Fig. 3.11 showing the RX_busy signal being asserted when synchronizer in clock2 domain is capturing the pulse of inA.

Example 3.7 Verilog testbench to simulate handshake signal from receive domain to transmit domain

```
module cdc3_tb ();
reg inA;
reg clock1;
reg clock2;
reg reset;
wire outB;
reg [3:0] i,j,k,l;

initial
begin
        clock1 = 0;
        j = $random;
        #j;
        forever #10 clock1 = ~clock1; // 50 MHz
```

```
end

initial
begin
        clock2 = 0;
        k = $random;
        #k;
        forever #50 clock2 = ~clock2; // 10 MHz
end

initial
begin
        reset = 0;
        inA = 0;
        #100;
        reset = 1;
        #100;
        reset = 0;
        #100;
        #6;
        inA = 1;
        #20;
        inA = 0;
        #155;
        inA = 1;
        #20;
        inA = 0;
        #1000;
        $stop;
```

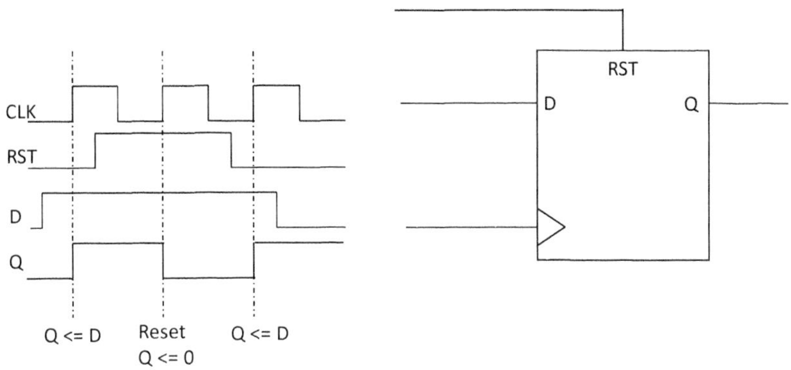

Fig. 3.12 Diagram showing synchronous reset flip-flop

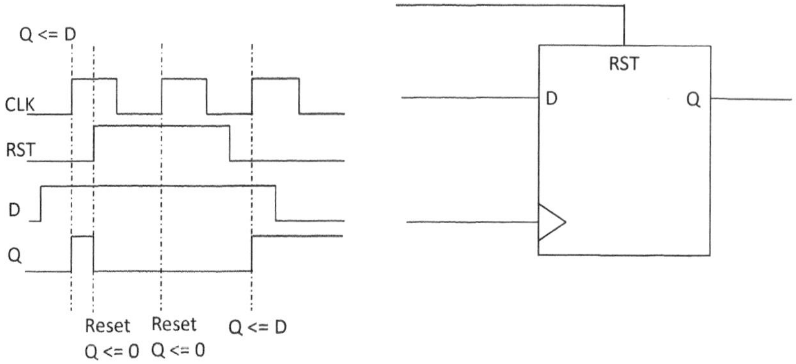

Fig. 3.13 Diagram showing asynchronous reset flip-flop

```
end

cdc3 cdc3_inst (.inA(inA), .outB(outB), .clock1(clock1), .
clock2(clock2), .reset(reset), .RX_busy(RX_busy));


endmodule
```

Referring to the simulation results shown in Fig. 3.11, the signal RX_busy can be used as an indication or qualifying signal for inA to ensure the pulses of inA are not too close to each other.

Reset

Reset is a common part of digital system and is present in every system. It allows a digital system to reset its logic and functionality. Reset can be divided into synchronous reset and asynchronous reset.

Synchronous reset is dependent upon clock, and the reset would only occur during a rising edge of clock for a rising edge-triggered flip-flop. Figure 3.12 shows the timing diagram for a synchronous reset flip-flop, and Example 3.8 shows the Verilog code for a synchronous reset flip-flop.

Example 3.8 Verilog code showing synchronous reset flip-flop

```
module syncreset (D, Q, RST, CLK);
input D, RST, CLK;
output Q;
```

```
reg Q;

always @ (posedge CLK)
begin
        if (RST)
                Q <= 0;
        else
                Q <= D;
end

endmodule
```

Referring to the Verilog code of Example 3.8, the sensitivity list of the always block "always @ (posedge CLK)" only consists of posedge CLK, which means that the always block is only triggered and evaluated when there is a rising edge of CLK.

Asynchronous reset can occur anytime irrespective of clock. A flip-flop with asynchronous reset can reset anytime irrespective of clock as long as the asynchronous reset signal is active (Fig. 3.13).

Example 3.9 Verilog code showing asynchronous reset flip-flop

```
module asyncreset (D, Q, RST, CLK);
input D, RST, CLK;
output Q;
reg Q;

always @ (posedge CLK or posedge RST)
begin
        if (RST)
                Q <= 0;
```

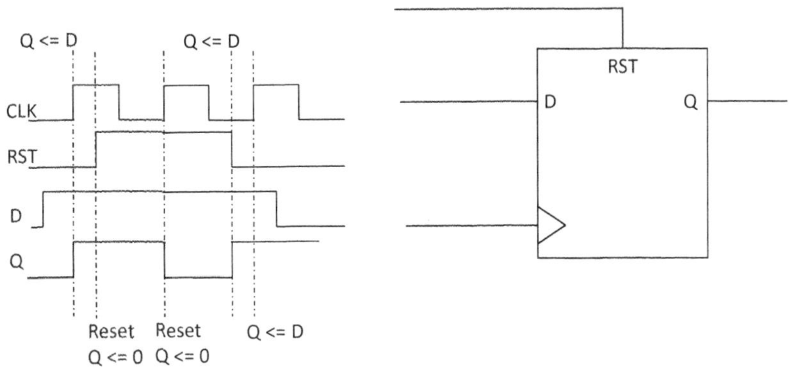

Fig. 3.14 Diagram showing waveform characteristic of wrongly coded asynchronous reset flip-flop

```
        else
                Q <= D;
end

endmodule
```

Referring to the Verilog code of an asynchronous reset flip-flop in Example 3.9, the sensitivity list of the Verilog code includes "posedge CLK or posedge RST." The always block is evaluated every time there is a positive edge detected on the CLK or the signal RST. Therefore, when RST changes from 0 to 1, irrespective of CLK, the sensitivity list of the always block in Example 3.9 is evaluated. Since RST signal is 1, the "if (RST)" statement is true, and Q is assigned 0.

An asynchronous flip-flop Verilog code is sometimes confused and written wrongly as shown in Example 3.10. This Verilog code although at a glance seems to represent the functionality of asynchronous reset, but upon detail analysis, the Verilog code is wrong and does not correctly represent the functionality.

Example 3.10 Wrongly coded Verilog for asynchronous reset

```
module wrongasyncreset (D, Q, RST, CLK);
input D, RST, CLK;
output Q;
reg Q;

always @ (posedge CLK or RST)
begin
        if (RST)
                Q <= 0;
        else
                Q <= D;
end

endmodule
```

Referring to Example 3.10, on the rising edge of RST when RST changes from 0 to 1, this will trigger the sensitivity list of "always @ (posedge CLK or RST)," and the always block is evaluated. On the rising edge of CLK, the sensitivity list is also triggered, and the always block is evaluated. This correctly corresponds to the functional behavior of the asynchronous reset flip-flop. However, the discrepancy happens during the falling edge of RST when RST changes from 1 to 0. This triggers the sensitivity list of "always @ (posedge CLK or RST)" and thereby causing evaluation of the always block. This causes the execution of statement "Q ≤ D" before rising edge of CLK, thus mismatch of the functionality with the correct behavior of the asynchronous reset flip-flop. Figure 3.14 shows the timing diagram for the wrongly coded asynchronous reset flip-flop. Notice the flip-flop output Q goes to logic 1 “Q ≤ D” on the falling edge of reset, thereby showing the wrong behavior of the asynchronous reset flip-flop. The code shown in Example 3.10 is also non synthesizable. Guideline for synthesizable code is explained in Chap. 8.

Chapter 4
Latch Inference

Latch inference refers to the condition whereby latches are inserted by synthesis tool to enable a signal to maintain its previous value. Inferring latches in a design is not desirable as it unnecessarily increases the size of the design. A bigger design will translate to a higher cost. A bigger design will also increase probability of defect and thereby reducing yield.

Apart from a larger design, inferring latches can also create timing issues. In digital design, it is a synchronous design whereby flip-flops are used as a timing path startpoint or timing path endpoint. As flip-flop only triggers its output change during rising edge of clock for a rising edge clock flip-flop, it is synchronous. An inferred latch is asynchronous as it is not dependent on the clock or clocked by a clock. Therefore, it can complicate timing analysis.

During synthesis, the synthesis tool will synthesize the RTL Verilog into logic gates. It is also during the synthesis process that the synthesis tool will insert inferred latches for those Verilog statement that does not have a complete if-else or case statement.

After synthesis, it is a common practice to make changes to the synthesized netlist to achieve testability. This is typically referred to as design for test (DFT). Chapter 5 discusses about DFT in detail. Due to requirements of DFT, latch inference complicates the process of making changes to a synthesized netlist to achieve testability.

Due to these complications, it is not recommended for digital design to have inferred latches. One common mistake in Verilog coding that causes latch inference is the use of incomplete if-else statement and incomplete case statement.

W. F. Lee, *Learning from VLSI Design Experience*,
https://doi.org/10.1007/978-3-030-03238-8_4

If-Else Statement

If-else statement is a procedural statement commonly used in always block to specify functionality. Most RTL Verilog code uses some form of if-else statement as it can easily describe required functionality of a digital circuit. However, incomplete usage of if-else statement can lead to latch inference. Example 4.1 shows a Verilog code using if-else statement but without specifying all the conditions of the if-else statement.

Example 4.1 Verilog code of if-else statement without specifying all conditions

```
always @ (inputA or inputB)
begin
        if (inputA & inputB)
                outputA = 0;
        else if (inputA & ~inputB)
                outputA = 1;
end
```

Referring to Example 4.1, the if-else statement specifies the conditions being compared for (inputA=1 and inputB=1) and (inputA=1 and inputB=0). It does not specify conditions for assignment of outputA if (inputA=0 and inputB=0) and (inputA=0 and inputB=1). During synthesis, the synthesis tool will insert a latch on the signal outputA such that it will maintain its previous value if the conditions of inputA and inputB are at logic values not specified in the if-else statement. This is referred to as latch inference.

Example 4.2 shows the same Verilog code of Example 4.1 but with an else condition added to include the unspecified conditions.

Example 4.2 Verilog code of if-else statement specifying all conditions

```
always @ (inputA or inputB)
begin
        if (inputA & inputB)
                outputA = 0;
        else if (inputA & ~inputB)
                outputA = 1;
        else
                outputA = 0;
end
```

Referring to Example 4.2, if inputA and inputB are at logic values not mentioned in the if-else statement, the functionality of else statement will be assigned to outputA.

Example 4.3 shows the same Verilog code as Example 4.2 but without using else statement. The functionality of Example 4.3 remains the same as Example 4.2 and without latch inference although no else statement is used. Example 4.3 specifies all possible combination of inputA and inputB, thus not needing an else statement to avoid the latch inference.

Example 4.3 Verilog code of if-else statement specifying all conditions

```
always @ (inputA or inputB)
begin
        if (inputA & inputB)
                outputA = 0;
        else if (inputA & ~inputB)
                outputA = 1;
        else if (~inputA & inputB)
                outputA = 0;
        else if (~inputA & ~inputB)
                outputA = 0;
end
```

When using if-else statement, it is recommended to always add the else statement to ensure there is no unwanted latch inference.

There is an exception in which an if-else statement without else statement does not result in a latch inference. This exception is in the coding of functionality of sequential logic, resulting in one or more flip-flop, as shown in Example 4.4.

Example 4.4 Verilog code of sequential logic

```
always @ (posedge clock or posedge reset)
begin
        if (reset)
                Q <= 0;
        else if (A & B)
                Q <= X1;
end
```

Referring to Example 4.4, an else statement is not required to avoid latch inference as the functionality of RTL code of Example 4.4 is sequential logic. When synthesized, the RTL code is translated to flip-flop for signal Q. The flip-flop acts as a register and is able to keep the previous value; thus, a latch will not be inferred.

Case Statement

Case statement is a procedural statement and is similar to if-else statement. Case statement is commonly used in always block to specify functionality. Most RTL Verilog code uses some form of case statement to describe functionality especially for state machine design. Incomplete usage of case statement can lead to latch inference. Example 4.5 shows a Verilog code using incomplete case statement that can cause latch inference.

Example 4.5 Verilog code of case statement without specifying all conditions

```
always @ (inputA or inputB or select)
begin
        case (select)
                2'b00: outputA = inputA;
                2'b01: outputA = inputB;
        endcase
end
```

Referring to Example 4.5, the case statement specifies the conditions of 2-bit select for "00" and "01." It does not specify conditions for assignment of outputA if select is "10" or "11." Therefore, if the unspecified conditions occur, the Verilog code of Example 4.5 will maintain the previous value of outputA. This means that outputA will be latched and create a condition of latch inference.

Example 4.6 shows the same Verilog code of Example 4.5 but with a default condition added to include the unspecified conditions.

Example 4.6 Verilog code of if-else statement specifying all conditions

```
always @ (inputA or inputB or select)
begin
        case (select)
                2'b00: outputA = inputA;
                2'b01: outputA = inputB;
                default: outputA = 0;
        endcase
end
```

Referring to Example 4.6, if the 2-bit select is at logic values not mentioned in the case statement such as "10" or "11," the functionality of default condition will be assigned to outputA.

Example 4.7 shows the same Verilog code as Example 4.6 but without using default condition. The functionality of Example 4.7 remains the same as Example 4.6 and without latch inference although no default condition is used. Example 4.7

specifies all possible combination of 2-bit select, thus not needing a default condition to avoid the latch inference.

Example 4.7 Verilog code of case statement specifying all conditions

```
always @ (inputA or inputB or select)
begin
        case (select)
                2'b00: outputA = inputA;
                2'b01: outputA = inputB;
                2'b10: outputA = 0;
                2'b11: outputA = 0;
        endcase
end
```

When using case statement, it is recommended to always add the default condition to ensure there is no unwanted latch inference.

Chapter 5
Design for Test

Digital circuits are large and commonly range in millions of gates. It is common to find ASIC chips or SoC chips with several million or more gates. Testing digital circuits for manufacturing defects can be a daunting task. Finding a manufacturing defect within the millions of gates and its millions of interconnects is a difficult and near impossible task without the help of proper testing methodology.

Manufacturing defects must be caught prior to the chip being delivered to customers as manufacturing defects can affect the functionality of the chip. Common manufacturing defects could be stuck at 0 fault and stuck at 1 fault. These faults could be caused by dust particles on the die during fabrication which causes a net to short to a 1 or short to a 0.

Figure 5.1 shows a diagram of a NAND gate in which one of the interconnect signal B shorted to VSS by a dust particle that may have drop onto the die during fabrication. This will create a "stuck-at-0" fault in signal B causing the output of the NAND gate to be always 1 irrespective of the logic value of input A. Such a fault is not a functional fault, but it can cause functional failures due to the NAND gate not able to switch logic values when input changes.

Apart from "stuck-at" faults, other faults that are commonly targeted such as transition faults and path delay faults are part of at-speed test. In at-speed tests, the design is operated at normal system clock speed to capture timing failures due to delay variations that may be caused by partially conducting transistor defects or resistive bridge defects.

Whether it's stuck at faults, at-speed test, or other manufacturing faults, the common method used to capture these manufacturing defects is using scan chain and ATPG (automatic test pattern generation). ATPG ensures the targeted defects are caught before the chip is shipped to customers.

W. F. Lee, *Learning from VLSI Design Experience*,
https://doi.org/10.1007/978-3-030-03238-8_5

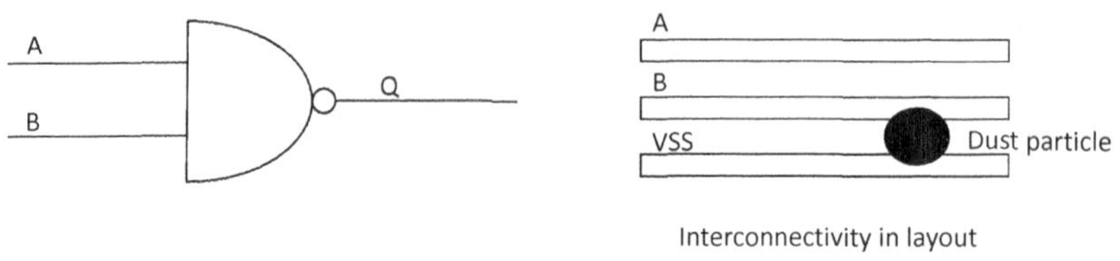

Fig. 5.1 Diagram showing metal connection B in layout shorted to VSS by a dust particle

Scan Chain

Scan chain is a method of testing where the flip-flops in a design are replaced with multiplexed flip-flops and connected in a series to form a scan chain. The input to the scan chain is a test input pin to the design called scan_in. The output of the scan chain is a test output pin from the design called scan_out. This process is referred to as test synthesis.

Referring to Fig. 5.2, during test synthesis, the data flip-flops are replaced with scan flip-flops which consist of a multiplexer in front of a data flip-flop. This is commonly referred to as scan cell replacement. Once the flip-flops are replaced with the scan flip-flops, they are stitched to form a scan chain. The input pin is the scan_in pin which can be shared with other functional input pin or a stand-alone scan_in pin. This scan_in pin is the beginning of the scan chain. The output of each scan flop is stitched to the mux input of the next scan flip-flop. The scan data is pumped into the design through this scan_in pin. The select of the multiplexer scan flip-flop is controlled by a scan_enable pin which allows toggling of the scan chain to either multiplex the scan data or combinational logic. This goes on from scan flip-flop to scan flip-flop until it reaches the end of the scan chain through scan_out pin. The scan_out pin can also be shared with a functional pin or can be a stand-alone scan_out pin.

Before Scan

The Verilog code shown in Example 5.1 is a simple RTL code of a design with three flip-flops, the input of the first flip-flop is a NAND function of inA and inB, the input of the second flip-flop is the NOR function of the output of the first flip-flop and inB, and the input of the third flip-flop is the XNOR function of the output of the second flip-flop and inB.

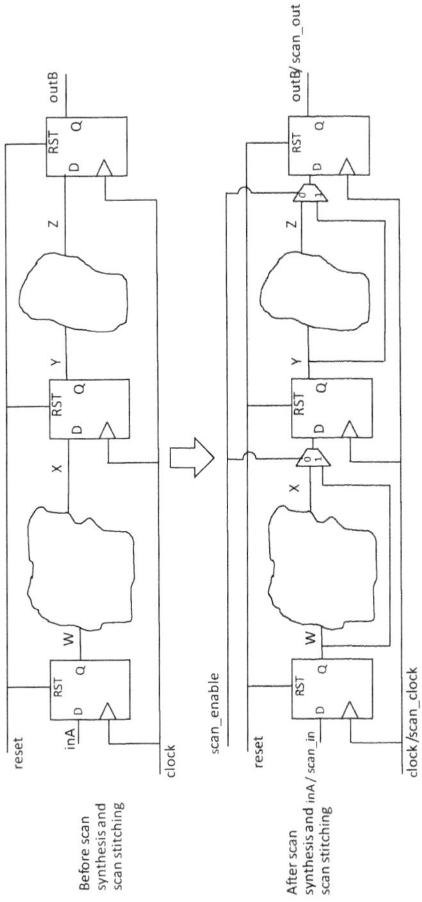

Fig. 5.2 Diagram showing before and after scan synthesis and scan stitching

Example 5.1 RTL Verilog code of a simple flip-flop design before scan

```
module beforescan (inA, inB, reset, clock, outB);
input inA, inB;
input reset, clock;
output outB;

reg inAq1, inAq2, outB;

always @ (posedge reset or posedge clock)
begin
       if (reset)
              begin
                     inAq1 <= 0;
                     inAq2 <= 0;
                     outB <= 0;
              end
       else
              begin
                     inAq1 <= ~(inA & inB);
                     inAq2 <= ~(inAq1 | inB);
                     outB <= ~ (inAq2 ^ inB);
              end
end

endmodule
```

Figure 5.3 shows the schematic of the synthesized netlist of the design. The synthesis tool writes this synthesized netlist as a Verilog netlist as shown in Example 5.2.

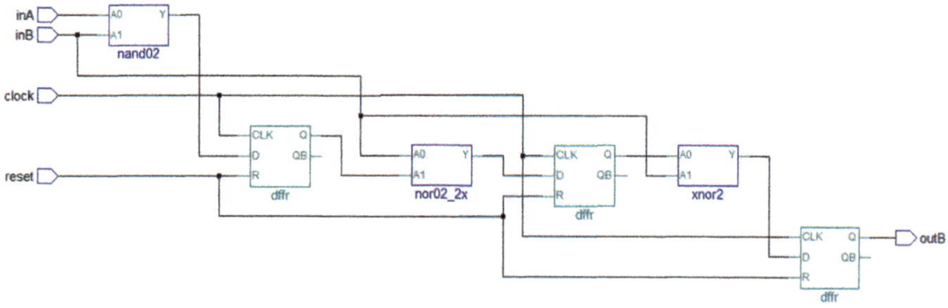

Fig. 5.3 Diagram showing synthesized netlist before scan

Example 5.2 Synthesized Verilog netlist of design before scan

```
module beforescan ( inA, inB, reset, clock, outB ) ;
    input inA ;
    input inB ;
    input reset ;
    input clock ;
    output outB ;
    wire inAq2, inAq1, nx4, nx14, NOT_nx20;
    wire [2:0] \$dummy ;
     dffr reg_outB (.Q (outB), .QB (\$dummy [0]), .D (NOT_nx20),
.CLK (clock), .R (reset)) ;
    xnor2 ix102 (.Y (NOT_nx20), .A0 (inAq2), .A1 (inB)) ;
    dffr reg_inAq2 (.Q (inAq2), .QB (\$dummy [1]), .D (nx14), .CLK
(clock), .R (reset)) ;
    nor02_2x ix15 (.Y (nx14), .A0 (inB), .A1 (inAq1)) ;
    dffr reg_inAq1 (.Q (inAq1), .QB (\$dummy [2]), .D (nx4), .CLK
(clock), .R (reset)) ;
    nand02 ix5 (.Y (nx4), .A0 (inA), .A1 (inB)) ;
endmodule
```

After Scan

During the scan synthesis process, the data flip-flops are replaced with scan flip-flops, and the scan input pins of the scan flip-flops are stitched to form a scan chain. Figure 5.4 shows the schematic after scan insertion and scan stitching is completed.

Referring to Fig. 5.4, the input pin scan_in is connected to the scan_in port of the first scan flip-flop in the scan chain. The output of the scan flip-flop connects to the scan_in port of the second scan flip-flop, and the output of the second scan flip-flop

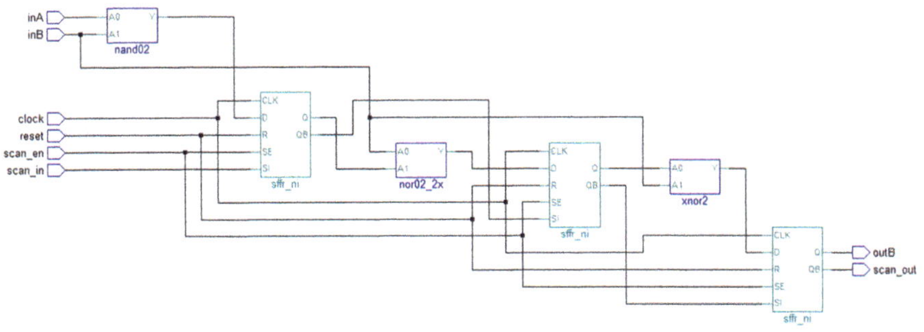

Fig. 5.4 Diagram showing schematic after scan

connects to the scan_in port of the third scan flip-flop. The output of the third scan flip-flop connects to the output scan_out, forming a scan chain from scan_in input to scan_out output.

Scan_en input controls the scan flip-flops to determine if the flip-flop passes the scan pin information or the data logic information. If the scan_en pin is low which means the design operates in functional mode, the flip-flop flops the data pin to the flip-flop output pin during rising edge of clock. If scan_en pin is high, this represents the design operating in scan mode. The logic values from the scan_in pin are passed through the scan chain on rising edge of every scan clock.

The scan synthesis tool can generate a scan Verilog netlist, as shown in Example 5.3, that is used for the next phase of design.

Example 5.3 Synthesized Verilog netlist of design after scan

```
module beforescan ( inA , inB , reset , clock , outB , scan_in ,
scan_out , scan_en );
input   inA , inB , reset , clock , scan_in , scan_en ;
output  outB , scan_out ;
wire inAq2 , inAq1 , nx4 , nx14 , NOT_nx20 ;
wire [2:0] \$dummy   ;

sffr_ni  reg_outB (.D ( NOT_nx20 ) , .SI ( \$dummy   [1] ) , .SE (
scan_en ) , .CLK ( clock ) , .R ( reset ) , .Q ( outB ) , .QB (
\$dummy   [0] ));
xnor2  ix102 (.A0 ( inAq2 ) , .A1 ( inB ) , .Y ( NOT_nx20 ));
sffr_ni  reg_inAq2 (.D ( nx14 ) , .SI ( \$dummy   [2] ) , .SE (
scan_en ) , .CLK ( clock ) , .R ( reset ) , .Q ( inAq2 ) , .QB (
\$dummy   [1] ));
nor02_2x  ix15 (.A0 ( inB ) , .A1 ( inAq1 ) , .Y ( nx14 ));
sffr_ni  reg_inAq1 (.D ( nx4 ) , .SI ( scan_in ) , .SE ( scan_en )
, .CLK  ( clock ) , .R ( reset ) , .Q ( inAq1 ) , .QB ( \$dummy
[2] ));
nand02  ix5 (.A0 ( inA ) , .A1 ( inB ) , .Y ( nx4 ));
assign  scan_out = \$dummy   [0] ;
endmodule
```

Automatic Test Pattern Generation (ATPG)

When scan chain is completed for a design, the next step is to generate the test pattern for the scan chain. This phase is termed ATPG (automatic test pattern generation). In this phase, an ATPG tool will analyze the functionality of the design and create a set of test patterns that can be used to identify defects in a design in manufacturing environment. For example, a common manufacturing defect that must be checked is "stuck-at" faults. It can be a "stuck-at-0" fault or "stuck-at-1" fault. A

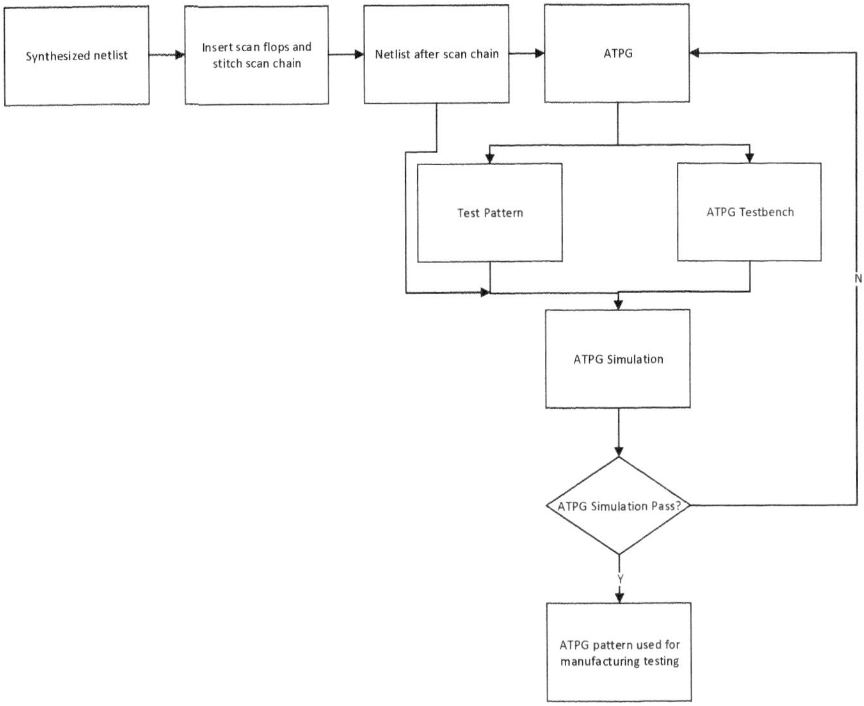

Fig. 5.5 Diagram showing flow for generation of ATPG pattern

"stuck-at-0" fault is caused by manufacturing defect that causes a signal stuck at 0, and a "stuck-at-1" fault is due to the signal stuck at 1. Other faults that are commonly targeted are transition faults and path delay faults which are part of at-speed test.

Whichever the faults targeted, the flow is similar. During ATPG, it generates a set of test patterns that is shifted into the design through the scan chain with scan_en enabled at logic 1. This will put the design in scan mode and the test patterns shifted into the scan chain. Scan_en is then deasserted, and the design goes into functional mode. The preloaded patterns will propagate through the design functionality and causes a logic value at scan_out. This value is checked against the expected pattern, and if a mismatch is observed at the scan_out pin, there is a manufacturing defect in the design.

Figure 5.5 shows the flow diagram for scan netlist and ATPG generation to obtain the ATPG test patterns that can be used by the manufacturing testing team to capture manufacturing defects.

Test Compression

For large ASIC design with thousands of flip-flops, creating a single scan chain with a single scan_in and scan_out may not be practical. A long scan chain will require many clock cycles during scan_en to load in the necessary scan values at each

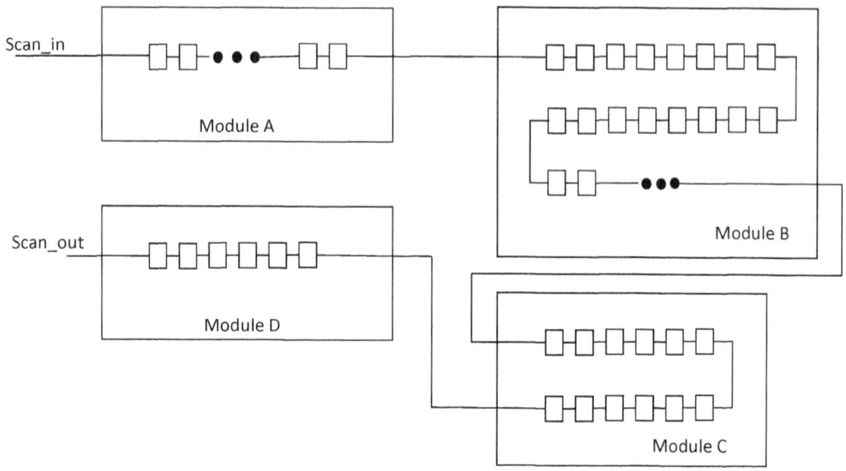

Fig. 5.6 Diagram showing fullchip with long scan chain

flip-flop along the scan chain. Figure 5.6 shows a fullchip with several modules that have many scan flip-flops forming a long scan chain.

A long scan chain will require many clock cycles to shift in a scan pattern with scan_en high. When the shift in of the scan pattern completes, scan_en goes low, and the loaded scan pattern at each scan flip-flop will propagate through the functional logic and produce an output which is compared by the tester to ensure the expected output of the fullchip matches the pattern. Another set of pattern is then shifted in again with scan_en high, followed by scan_en low for the pattern to propagate through the functional logic and compared for matching between the fullchip output and expected output. This repeats itself for entire set of ATPG patterns. Many clock cycles are required for a design with a long scan chain.

A design with a long scan chain that is being tested in manufacturing will require many cycles per pattern, and with many patterns to test the design, the tester will need a lot of time to test the design. This increases testing cost dramatically as longer testing time results in less chips being tested, thus higher test cost per chip.

Design with long scan chain can be broken up into shorter scan chain forming multiple scan chain as shown in Fig. 5.7. Having multiple scan chain can shorten scan pattern shift in time when scan_en is high. However multiple scan chain requires the tester to drive multiple scan_in pins at the tester level which reduces test time but increases test cost as less chips can be tested in parallel by a tester.

Let's consider a tester in production can drive/monitor 100 pins. For simplicity of understanding, let's consider only scan_in and scan_out pins for this 100 pins that the tester can drive/monitor, assuming the tester can drive additional pins for scan_en and scan_clock. A design with a single long scan chain can be tested 50 chips in parallel by the tester. However each pattern shift in time is long and increases test time. A design with five scan chains (five scan_in pins and five scan_out pins, scan_en, and scan_clock are the same for the five scan chains) can be tested with ten chips in parallel by the tester. The test time is reduced with multiple

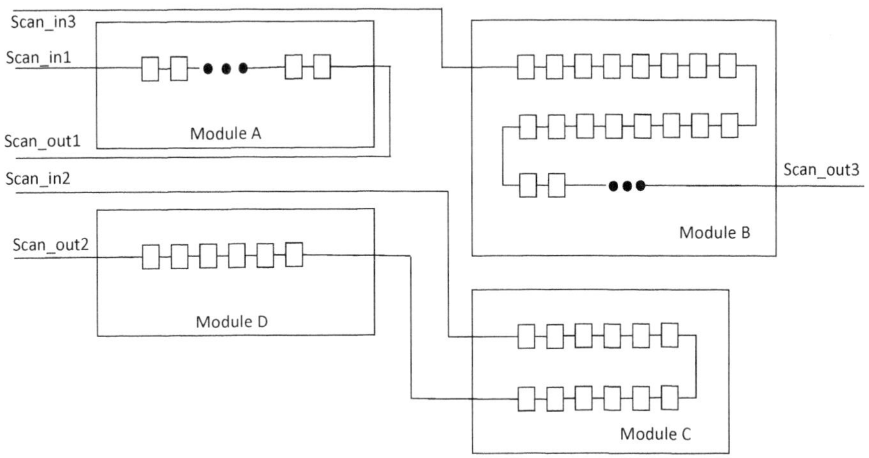

Fig. 5.7 Diagram showing fullchip with multiple scan chain

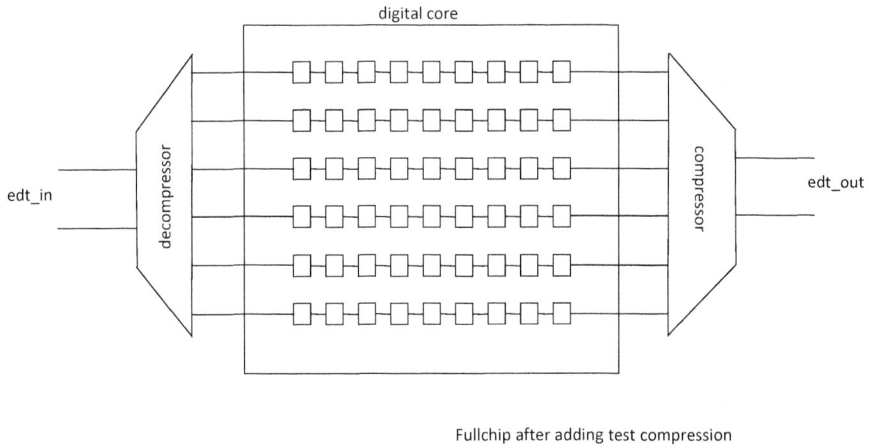

Fig. 5.8 Diagram showing fullchip after adding test compression

scan chains, but the amount of chips that a tester can test in parallel is reduced as each chip requires more test pins for the multiple scan chains.

Both methods of a single long scan chain or multiple shorter scan chains result in increase of test cost. A more efficient solution to reducing test cost with having multiple scan chain and less test pins to be driven/monitor by tester is test compression.

In test compression, multiple scan chains are created instead of a single long scan chain. Each of these scan chain have its scan_in and scan_out signals creating multiple parallel scan chains. Two additional logic blocks are added to the design to decompress and to compress the ATPG patterns, as shown in Fig. 5.8.

Referring to Fig. 5.8, the digital core is the digital design where the flip-flops are stitched to form multiple scan chains. This will ensure each scan chain is not overly

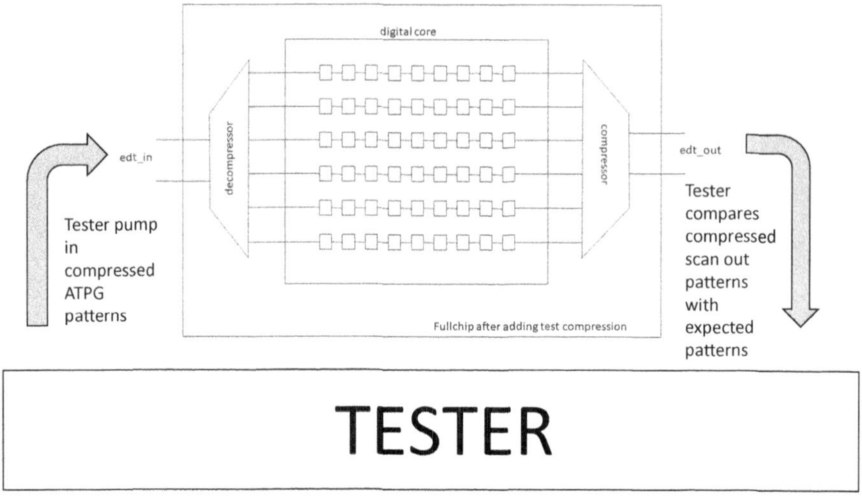

Fig. 5.9 Diagram showing tester interaction with fullchip

long when compared to using only one scan chain. During test compression, two digital blocks are added to the digital core to form a larger fullchip which consists of the decompressor block, the digital core with its multiple scan chains, and the compressor block. The compressed ATPG patterns are generated by the ATPG compression tool. It is similar to the ATPG pattern except that it is compressed in order to obtain a smaller ATPG pattern size. This compressed ATPG pattern will be used by the tester during manufacturing process to determine if there's any manufacturing defects.

The decompressor block added to the digital core on the fullchip is a logic block to decompress the compressed ATPG patterns and shifted into the corresponding scan chains. The output from each scan chain in the digital core is compressed through the compressor logic block to obtain a compressed scan output. This compressed scan output is compared to an expected compressed ATPG pattern on a tester to determine if there are any failures. Figure 5.9 shows a diagram on the interaction between the fullchip and the tester.

It is important that the insertion of the decompressor block and compressor block to the digital core to form the fullchip must not affect the digital core's functional timing. The decompressor block and compressor block are only used during ATPG testing and are disabled during normal functional operation of the chip.

Scan Chain Crossing Different Clock Domains During Shift Phase of ATPG

When a design consists of two different asynchronous clock domains, a signal crossing between one asynchronous clock and the other asynchronous clock will result in metastability issues. To work around this problem, synchronizers were introduced to resolve these CDC (clock domain crossing) issues. CDC issues affect the functionality of the design, and synchronizers can help increase the MTBF (mean time between failure) of the metastability when crossing asynchronous clock domain.

Another issue that affects design when crossing clock domain is during the shift phase of a scan chain. In DFT, circuits with different clock domains are stitched to the same scan chain with the last flip-flop of the first clock domain connected to the first flip-flop of the second clock domain, as shown in Fig. 5.10.

As clock1 tree and clock2 tree are two separate clock trees, their clock skew varies greatly between them. When scan chain crosses from one clock domain to another clock domain, timing becomes critical during the shift phase of ATPG. During scan shift of ATPG, scan_enable is at logic 1, and scan clock shifts in the necessary preconfigured ATPG values to the scan flip-flops. These values are shifted in from scan_in through the scan flip-flops. When the shifting crosses from clock1 domain to clock2 domain, feedthrough can occur as shown in Fig. 5.11.

Referring to Fig. 5.11, when scan_enable is at logic 1, scan_clk is connected to both clock1 and clock2. However, due to the difference in clock skew between the two clock domains, scan flip-flop X flops the shift data to net1 during rising edge of scan_clk (net1 is valid t1 ns after rising edge of clock1). As clock2 tree have a larger clock skew, flip-flop Y only observes the rising edge of scan_clk at a later time. Flip-flop Y will flop the shift_in data from net1 to net2 t2 ns after rising edge of clock2. This situation is undesirable as the shift data is shifted through two scan flip-flops within one cycle of scan_clk. This results in a feedthrough situation when the scan data is shifted from the last flip-flop of clock1 domain to the first flip-flop of clock2 domain within one scan_clk cycle.

A latch can be introduced between the two clock domains to resolve the feedthrough problem that occur when both clock domains are connected to scan_ clk. Figure 5.12 shows the same scan chain of Fig. 5.11 with a latch between the last scan flip-flop of clock1 domain and the first scan flip-flop of clock2 domain.

Referring to Fig. 5.12, a latch is connected to the inverse of clock1 (inverse of scan_clk). When rising edge of scan_clk, net1 is not valid immediately with the flopped data by scan flip-flop X. The inverse of scan_clk becomes high and puts the latch into transparent mode. The scan data flopped by scan flip-flop X is latched by the latch, and net1 becomes valid t3 ns after fall of scan_clk. Scan flip-flop Y will flop the scan data at the next rising edge of clock2 (scan_clk), and net2 is valid t4 ns after rising edge of next clock2 (scan_clk). The latch is referred to as a lockup latch and prevents the feedthrough condition during the shift phase of ATPG.

Example 5.4 and Fig. 5.13 show the netlist and schematic of a design consisting of two flip-flops, one clocked by clock1 and the other by clock2.

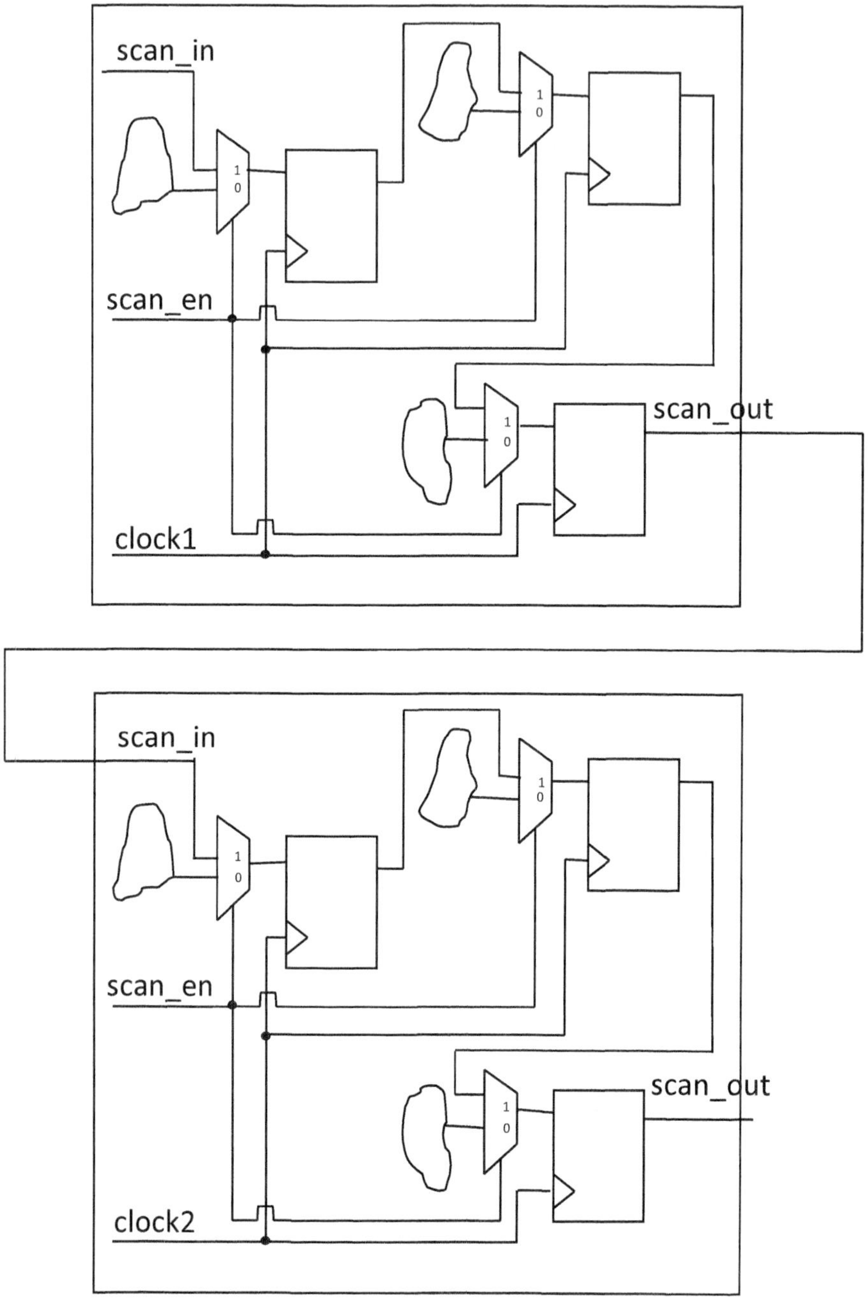

Fig. 5.10 Diagram showing scan chain crossing clock domain

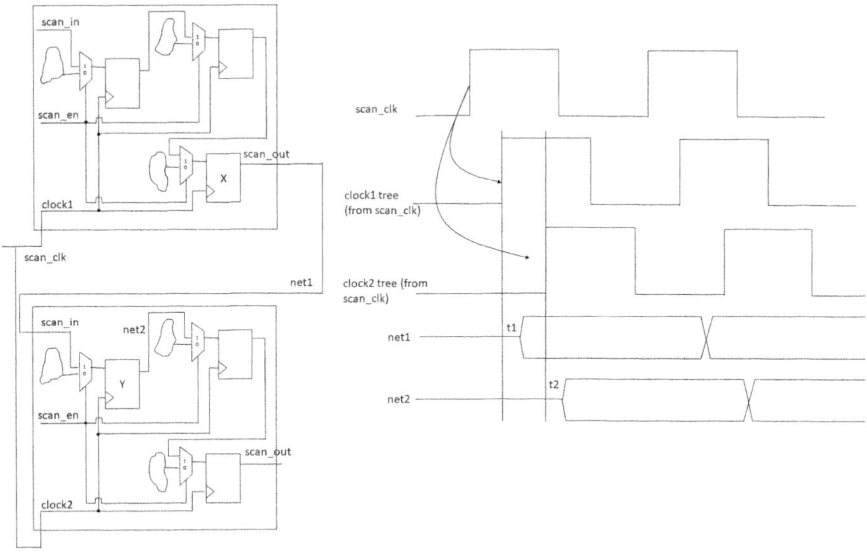

Fig. 5.11 Diagram showing scan chain crossing clock domain resulting in feedthrough

Example 5.4 Gate level netlist with two functional clocks

```
module cdcscan ( inA, inB, clock1, inC, clock2, reset, net2 ) ;

    input inA ;
    input inB ;
    input clock1 ;
    input inC ;
    input clock2 ;
    input reset ;
    output net2 ;

    wire net1, nx0, nx6;
     wire [1:0] \$dummy ;

     dffr reg_net2 (.Q (net2), .QB (\$dummy [0]), .D (nx6), .CLK
(clock2), .R (reset)) ;
    or02 ix7 (.Y (nx6), .A0 (net1), .A1 (inC)) ;
     dffr reg_net1 (.Q (net1), .QB (\$dummy [1]), .D (nx0), .CLK
(clock1), .R (reset)) ;
    and02 ix1 (.Y (nx0), .A0 (inB), .A1 (inA)) ;
endmodule
```

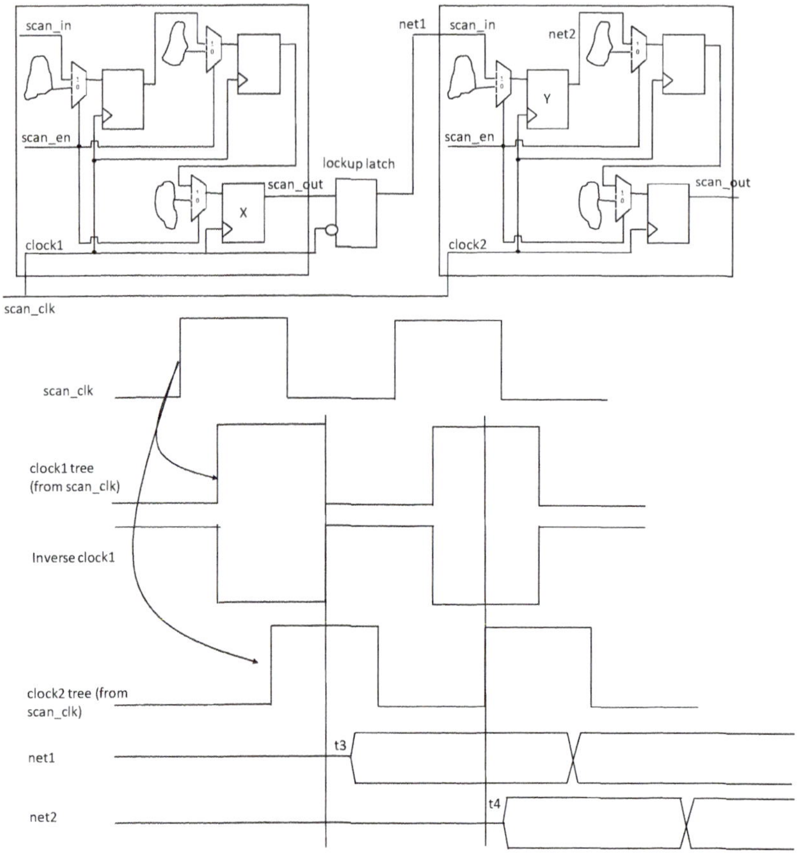

Fig. 5.12 Diagram showing clock1 and clock2 domain connected to scan_clk with a latch between the two clock domains

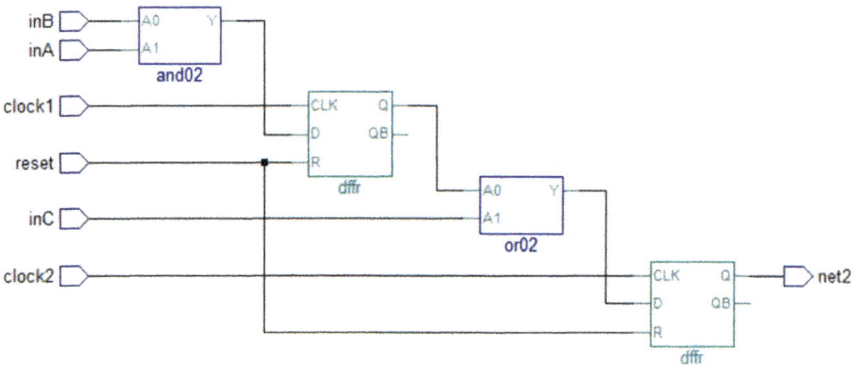

Fig. 5.13 Diagram showing schematic of Example 5.4

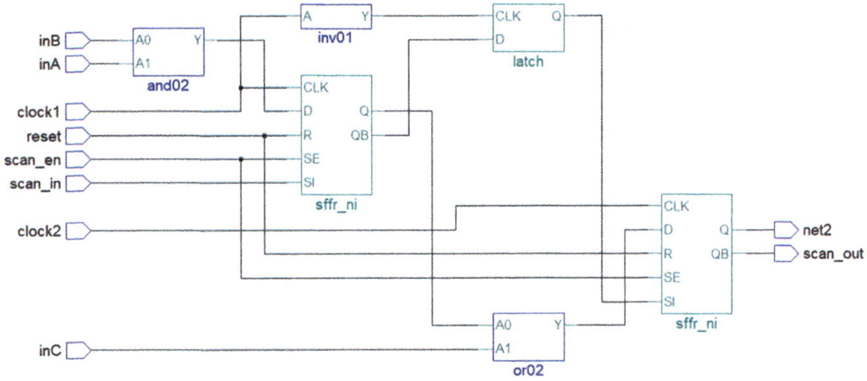

Fig. 5.14 Diagram showing schematic of Example 5.5 with lockup latch

The gate level netlist of Example 5.4 is put through a scan insertion and scan chain stitching using Mentor Graphics' DFTAdvisor. Example 5.5 shows the gate level netlist after scan insertion and stitching. Figure 5.14 shows the schematic after scan insertion and stitching.

Example 5.5 Gate level netlist with two functional clocks after scan insertion and stitching

```
module cdcscan ( inA , inB , clock1 , inC , clock2 , reset , net2
, scan_in ,
                  scan_out , scan_en );
input   inA , inB , clock1 , inC , clock2 , reset , scan_in ,
scan_en ;
output  net2 , scan_out ;
  wire net4 , net3 , net1 , nx0 , nx6 ;
  wire [1:0] \$dummy   ;

  inv01  uu1 (.A ( clock1 ) , .Y ( net4 ));
  latch  lckup1 (.D ( \$dummy[1] ) , .CLK ( net4 ) , .Q ( net3 ));
  sffr_ni  reg_net2 (.D ( nx6 ) , .SI ( net3 ) , .SE ( scan_en ) ,
.CLK  ( clock2 ) , .R ( reset ) , .Q ( net2 ) , .QB ( \$dummy[0]));
  or02  ix7 (.A0 ( net1 ) , .A1  ( inC ) , .Y ( nx6 ));
  sffr_ni  reg_net1 (.D ( nx0 ) , .SI ( scan_in ) , .SE ( scan_en
) , .CLK ( clock1 ) , .R ( reset ) , .Q ( net1 ) , .QB ( \$dummy[1]
));
  and02  ix1 (.A0 ( inB ) , .A1  ( inA ) , .Y ( nx0 ));
  assign  scan_out = \$dummy[0] ;
endmodule
```

Referring to Fig. 5.14, a lockup latch is connected between the flip-flop reg_net1 and reg_net2. The lockup latch is connected to the transmitting clock which in this case is clock1. The lockup latch will ensure the shifted scan data will not feedthrough from the last scan flip-flop on clock1 domain to the first scan flip-flop on clock2 domain.

An alternative to using a lockup latch to fix the difference of clock skew between clock1 and clock2 is by inserting buffers to clock1 domain to balance the two clock trees, as shown in Fig. 5.15. The buffers inserted in clock1 domain will balance the clock1 tree with clock2 tree thereby not requiring a lockup latch. However this method is not commonly used as the difference in clock skew between clock1 tree and clock2 tree may require a large number of buffers. Furthermore the buffers will have variation in delay across PVT (process, voltage, and temperature) thereby causing the two trees to balance under certain PVT corners but fail in others.

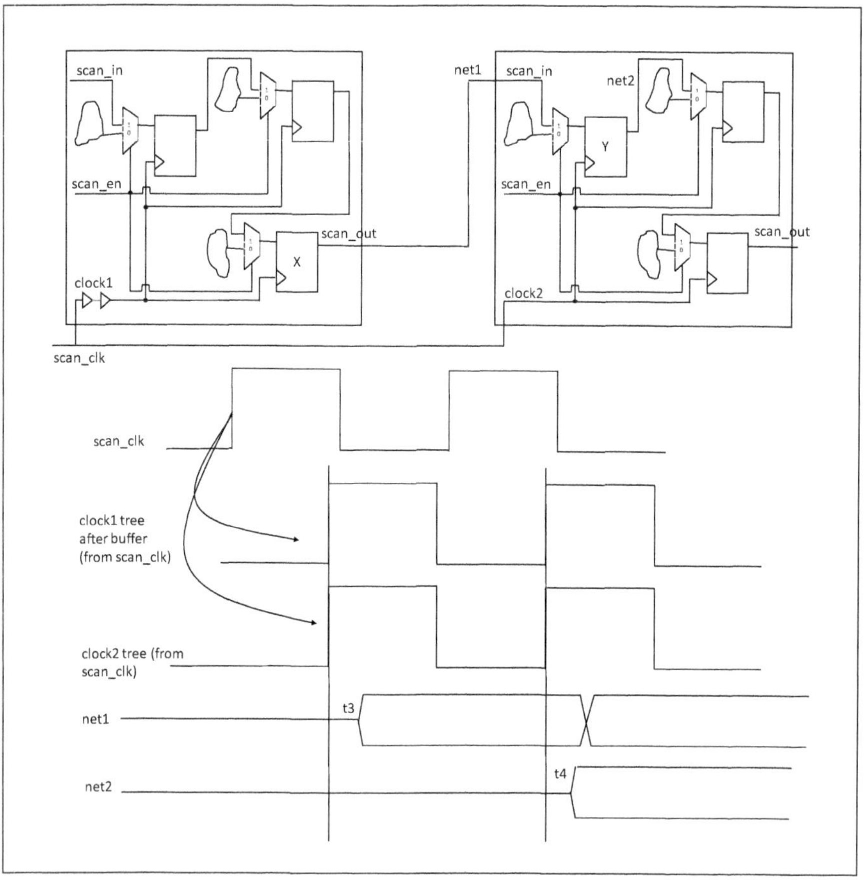

Fig. 5.15 Diagram showing buffer insertion instead of lockup latch at clock1 to fix scan shift feedthrough

Scan Chain for Design with Different Power Domains

Many of today's ASIC and SOC have low power requirement resulting in the chip having several different power domains. These power domains operate in different voltages, and its corresponding logic resides within its power domain island. When a functional signal crosses from one power domain to another power domain, a level shifter is needed to bridge the signal between the two different power domains, as shown in Fig. 5.16.

When a signal crosses from one power domain to another power domain, the signal may become invalid and nonfunctional due to the difference in voltage level. Such a signal must be bridged between the two power domains by a level shifter. The level shifter acts as a cell to bring the signal from the signal source voltage level to the signal destination voltage level.

During stitching of scan cells to form scan chain, the designer must be aware to minimize the crossing of scan chain from one power domain to another power domain as each crossing of the chain will require the insertion of a level shifter. The level shifter increases area and increases delay, and minimizing the crossing of the scan chain between power domains is desirable. To avoid such a scenario, one possible solution is to create an individual scan chain for each power domain. This ensures the scan chain does not cross power domain and removes the necessity of level shifters in the scan chain, but more scan pins are needed. Figure 5.17 shows a design with two power domains. One method is to limit the scan chain to the power domain, thus one scan chain per power domain. This is the simplest solution but requires more scan pins. The second method is to combine the power domain into a scan chain, but care must be taken to consider level shifters when the scan chain crosses power domain.

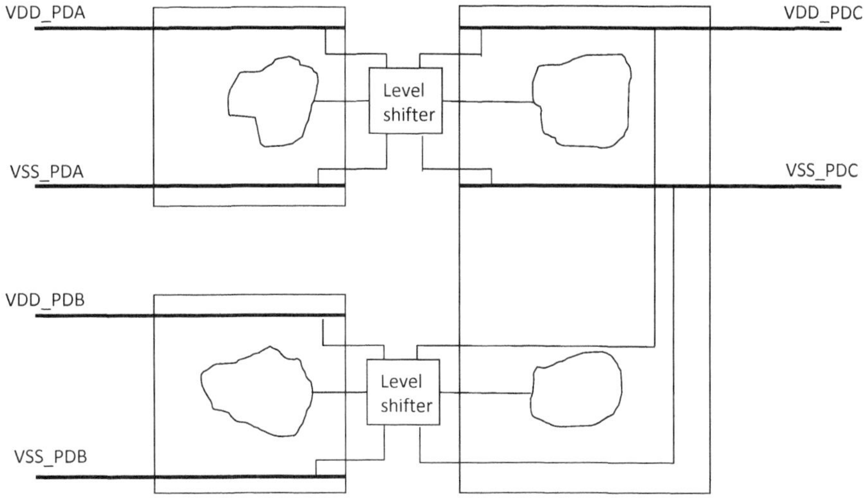

Fig. 5.16 Diagram showing signal crossing power domain requiring level shifter

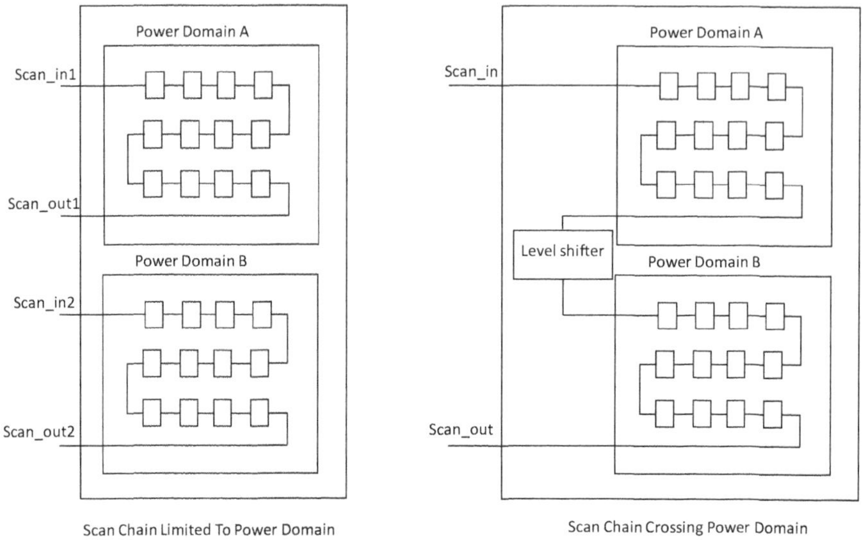

Fig. 5.17 Diagram showing two different methods of handling scan chain with power domain

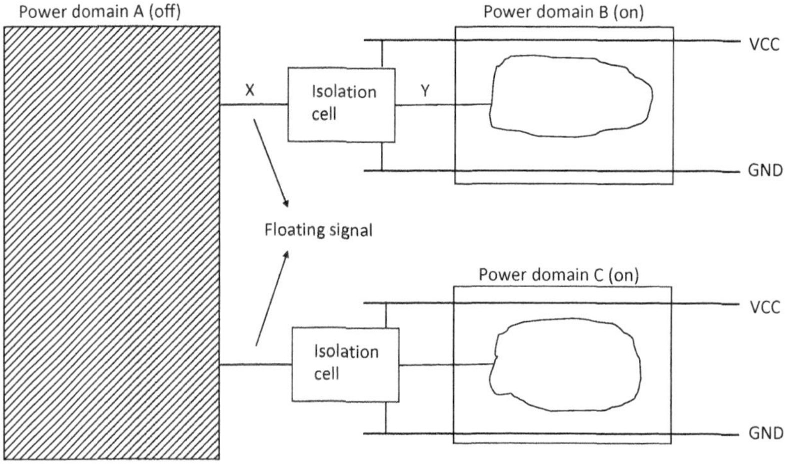

Fig. 5.18 Diagram showing isolation cell bridging an off power domain to an on power domain

The use of level shifters when the scan chain crosses power domain is applicable if the power domains are on during testing. If one of the power domains is off during testing, an isolation cell is required instead of a level shifter.

When a functional signal crosses from a power domain to another, when the transmitting signal power domain switches off, the transmitting signal voltage level floats, and the functionality of the receiving signal will be affected. For such conditions, an isolation cell is required as shown in Fig. 5.18. The isolation cell will tie the floating signal X to logic signal Y which can be a logic 1, logic 0, or previous latched value of signal X, depending on the type of isolation cell used.

A scan chain that crosses a power domain will require an isolation cell at the boundary of the power domain it crosses if the source power domain is off during certain test conditions.

Capture Phase of ATPG for Multiple Clock Design

ATPG is divided into shift phase and capture phase. When scan shifting of ATPG across clock domain, feedthrough may occur resulting in the shift data crossing the last scan flip-flop of clock1 domain to the first flip-flop of clock2 domain. To avoid this problem, a lockup latch is introduced between the two clock domains as discussed in Sect. 5.5.

During the second stage of ATPG which is the capture stage, scan_enable is kept low, and the shifted ATPG data will propagate through the functional path as shown in Fig. 5.19. A lockup latch cannot be used during capture stage of ATPG to work around the difference in clock skew between the clock domains because capture stage involves the functional path. Introducing a lockup latch into the functional path during capture will affect the timing of the functional path and will cause functional timing failures. The work-around to the capture stage is by clocking each clock separately.

Referring to Fig. 5.19, scan_enable is low; output from the first scan flip-flop will propagate through the functional path of combinational logic instead of lockup latch. As the functional path involves two separate clocks, ATPG will clock each clock separately. Using the gate level netlist of Example 5.5 which have a lockup latch inserted between the two scan flip-flops crossing clock1 domain to clock2 domain, the simulation of the generated ATPG pattern is shown in Fig. 5.20.

Referring to Fig. 5.20, the ATPG pattern is being shifted into the design when scan_enable is 1.

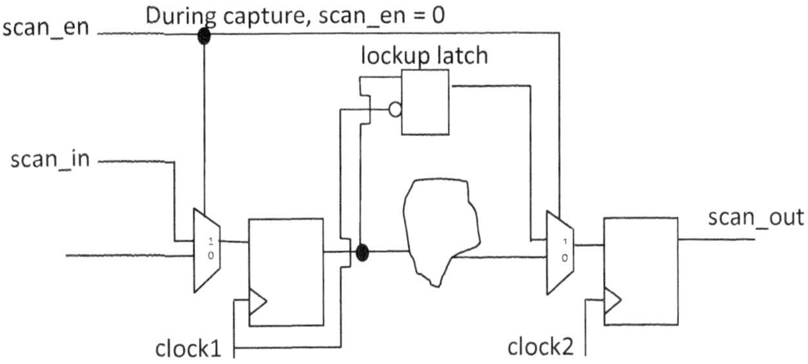

Fig. 5.19 Diagram showing capture stage of ATPG with the shifted ATPG data propagating through functional path

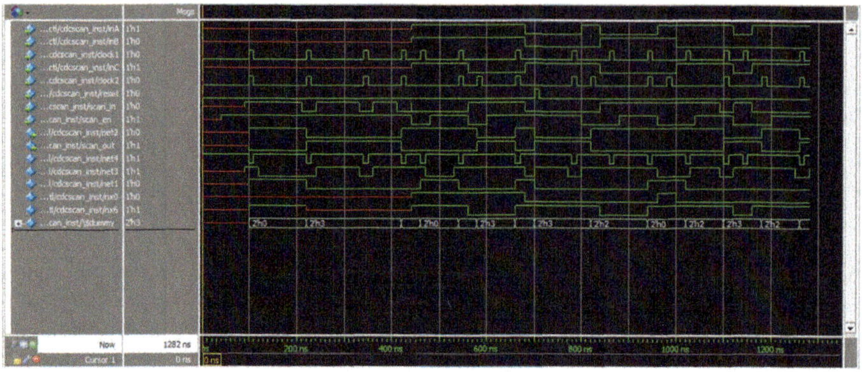

Fig. 5.20 Diagram showing simulation of ATPG pattern of Example 5.5

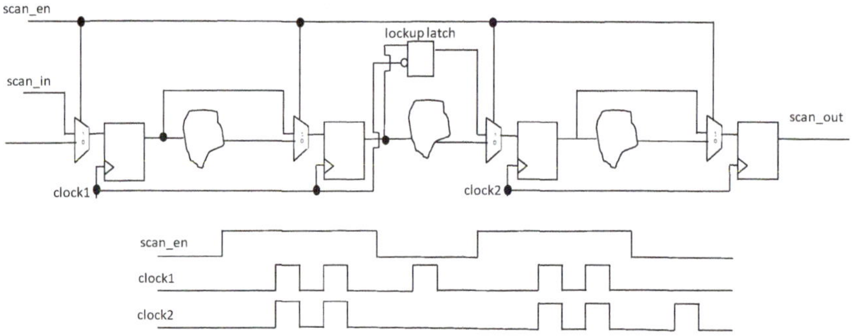

Fig. 5.21 Diagram showing capture stage in ATPG clocking clocks separately

- From time 40 ns to 440 ns, ATPG data is shifted in through scan_in with clock1 and clock2 being pulsed together as scan_clk.
- From time 440 ns to 480 ns, scan_en is 0, and clock1 is pulsed for capture.
- From time 480 ns to 560 ns, scan_en is 1, and clock1 and clock2 are pulsed together as scan_clk to shift in ATPG data.
- From time 560 ns to 600 ns, scan_en is 0, and clock2 is pulsed for capture.
- The shift and capture are repeated for other ATPG patterns.

From the simulation waveform shown in Fig. 5.20, the capture mode only pulses one clock at a time during capture mode, although both clocks are pulsed together during the shift mode.

In general, during ATPG, shift mode will pulse the clocks together as scan_clk with a lockup latch used between different clock domains. During capture, each clock is pulsed separately instead of using lockup latch as capture stage involves functional path. Using a lockup latch in functional path will affect the functional path timing.

Figure 5.21 shows two clock designs with each clock domain having two scan flip-flops. The ATPG shift phase will have two shift clocks with both clock1 and clock2 being shifted together. The ATPG capture phase will drive the two clocks separately with clock1 driven first then followed by clock2.

Logic Built in Self-Test

When testing a chip during manufacturing using ATPG, a pre-generated ATPG pattern is pumped into the chip by a tester and the output of the chip monitored by the tester and compared to an expected output. When the output matches the expected pattern, the chip passes ATPG test. When the output does not match the expected pattern, the chip fails the ATPG test. As the ATPG patterns gets very large for complex large ASIC chips, test compression is used whereby the patterns are compressed. But a tester is still required to pump in the compressed pattern into the chip, and the tester will match the output of the chip with a compressed expected pattern to determine if the chip passes the ATPG test.

However there may be cases where a chip may not be easily accessible to expensive external testers and therefore not suitable for ATPG testing. There could also be critical applications of a chip, for example, a chip used in automotive for braking system. Such a chip's functionality is critical to the automotive's braking system and must be tested every time the car is started. The logic BIST circuitry inside the chip will automatically run a self-test when the car is started. Any failure to the logic BIST self-test will have the car electronics alert the driver. Other markets where the chips must be fully functional such as military, medical, automotive, and aerospace commonly use logic BIST for system self-tests to ensure the chips are in full functionality.

How Does Logic BIST Work

Logic BIST achieves logic circuit self-test by generating pseudorandom patterns and pumping these patterns into scan chain. A multiple input signature register (MISR) collects the outputs of the scan chain and generates a corresponding MISR signature. This MISR signature is then read out of the chip serially and compared externally to the chip to determine if the logic BIST passes or fails.

Figure 5.22 shows a top level block diagram of a logic BIST consisting of the following components:

(a) BIST controller

1. The main controller circuit which controls the PRPG (pseudorandom pattern generator).
2. Sends control signals to PRPG to generate the pseudorandom patterns.

(b) PRPG

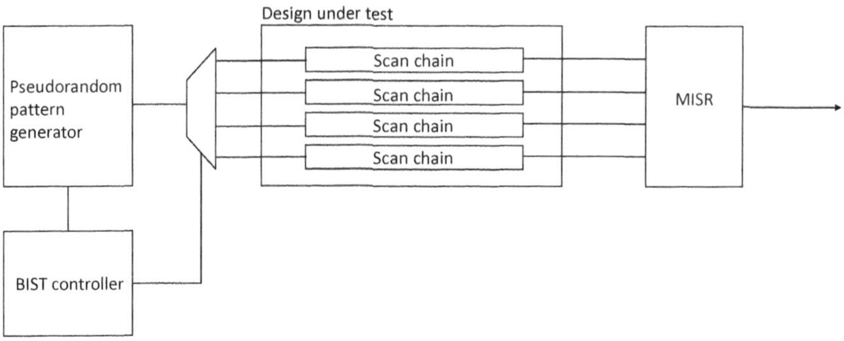

Fig. 5.22 Diagram showing block diagram implementation of logic BIST

1. The pseudorandom pattern generator that generates the necessary pseudorandom patterns to be pumped into the scan chain.

(c) Scan chain

1. The pseudorandom patterns are pumped into the scan chain.
2. The logic output is feed into a MISR.

(d) MISR (multiple input signature register)

1. MISR generates a signature, and this signature is serially shifted out of the chip for comparison externally.
2. A match in the MISR signature means the logic BIST passes.

Referring to Fig. 5.22, the implementation of logic BIST increases circuit size; however, it does not require external tester and reduces test cost while increasing die cost.

Implementation of Logic BIST

Implementation of logic BIST requires scan chain as ATPG does. Using the scan chain design of Example 5.5 which is a design of two flip-flops driven by clock1 and clock2 with a lockup latch between them, logic BIST can be implemented for the design. Using Mentor Graphics's logic BIST, the gate level netlist of Example 5.5 is put through the logic BIST flow as shown in Fig. 5.23.

Different design tools have different logic BIST implementation flows, but the end result remains the same with the insertion of logic BIST circuitry into the design to allow logic self-test. Figure 5.24 shows the top level block diagram after implementation of logic BIST of Example 5.5.

During logic BIST implementation, a series of additional pins will be added which is used to control the logic BIST. Figure 5.25 shows the schematic for the block cdcscan which is the original design with scan chain after logic BIST insertion.

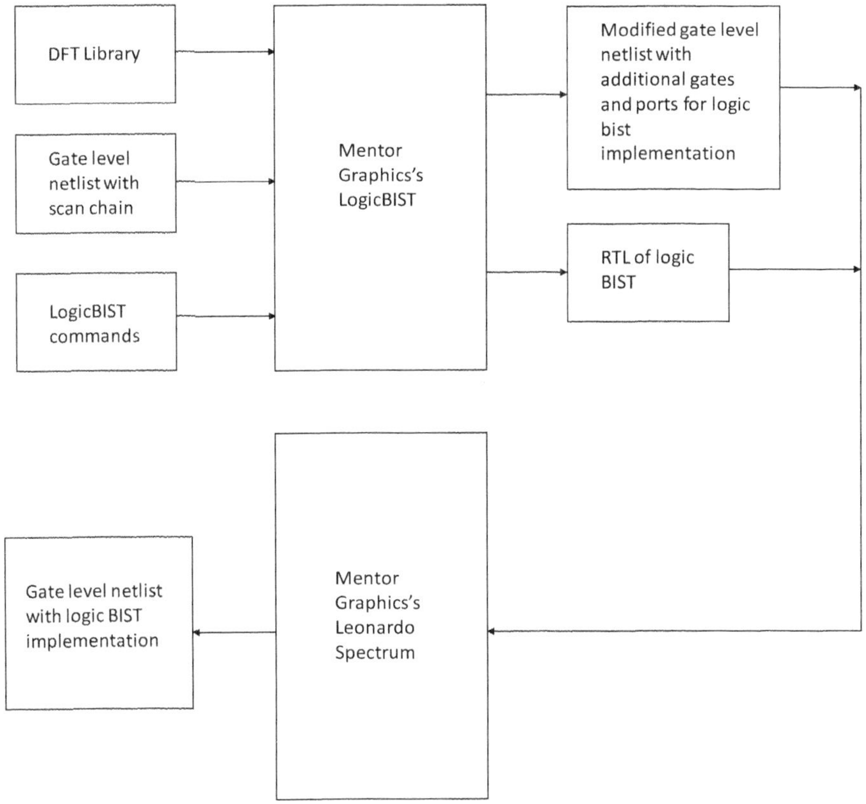

Fig. 5.23 Diagram showing implementation flow of logic BIST using Mentor Graphics's logic BIST and Leonardo Spectrum

Referring to Fig. 5.25, the scan chain with the lockup latch between the two scan flops controlled by clock1 and clock2 remains unchanged, but additional circuitry is added to cater to logic BIST.

Figure 5.26 shows the schematic for the logic BIST controller block with more subblocks underneath the BIST controller. The area report shown in Example 5.6 is generated from Mentor Graphics's Leonardo Spectrum synthesis tool. It shows the logic BIST controller consists of 1542 gates using TSMC 0.35 micron standard cell library through ADK (ASIC Design Kit), and the cdcscan consists of 34 gates. It can be concluded that logic BIST increases die size, but its advantages of being able to perform self-test on its logic are valuable especially for chips used for certain applications such as automotive and aerospace where the chip self-tests every time the system is switched on. Any logic BIST self-test failure will be caught, and the system user alerted to the failure.

Fig. 5.24 Diagram showing top level block diagram after logic BIST implementation

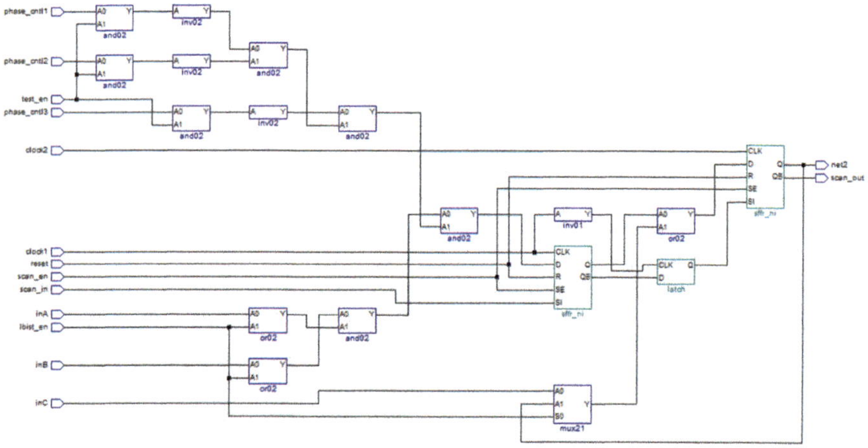

Fig. 5.25 Diagram showing schematic for module cdcscan after logic BIST insertion

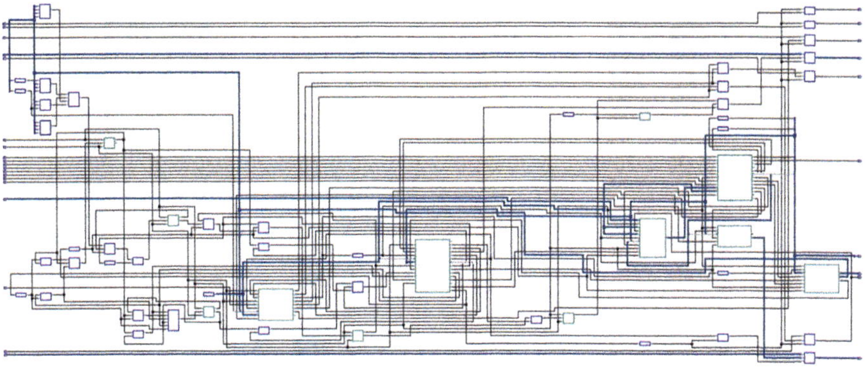

Fig. 5.26 Diagram showing schematic for logic BIST controller

Example 5.6 Area report on cell usage of top level after logic BIST insertion

```
*******************************************************
Cell:   pc_pattern_count_2_4_4_15_000000000000000_111111111111111
View: INTERFACE    Library: work
*******************************************************
 Cell         Library   References     Total  Area
and03         tsmc035_typ     3 x        2        5 gates
aoi21         tsmc035_typ     2 x        1        2 gates
aoi33         tsmc035_typ     6 x        2       13 gates
buf02         tsmc035_typ     5 x        1        5 gates
dffr          tsmc035_typ    18 x        6      100 gates
dffs_ni       tsmc035_typ    15 x        6       87 gates
inv01         tsmc035_typ    16 x        1       12 gates
inv02         tsmc035_typ     9 x        1        7 gates
mux21_ni      tsmc035_typ    26 x        2       47 gates
nand02        tsmc035_typ    15 x        1       15 gates
nand03        tsmc035_typ     2 x        1        2 gates
nand04        tsmc035_typ     4 x        1        6 gates
nor02_2x      tsmc035_typ     1 x        1        1 gates
nor02ii       tsmc035_typ    12 x        1       15 gates
nor04_2x      tsmc035_typ     1 x        2        2 gates
oai21         tsmc035_typ    18 x        1       22 gates
oai22         tsmc035_typ     1 x        1        1 gates
or02          tsmc035_typ     5 x        1        6 gates
xnor2         tsmc035_typ    14 x        2       27 gates
xor2          tsmc035_typ     2 x        2        4 gates

 Number of ports :                       41
 Number of nets :                       199
```

```
 Number of instances :                    175
 Number of references to this view :        1

Total accumulated area :
 Number of gates :                        381
 Number of accumulated instances :        175

*****************************************************
Cell:  sm_scan_monitor_1_0_17_1110000000000000_1111111111111111
View: INTERFACE    Library: work
*****************************************************
 Cell        Library   References     Total Area
buf02        tsmc035_typ      3 x       1       3 gates
dffs_ni      tsmc035_typ     17 x       6      99 gates
inv01        tsmc035_typ      2 x       1       2 gates
inv02        tsmc035_typ      3 x       1       2 gates
latch        tsmc035_typ      1 x       2       2 gates
mux21_ni     tsmc035_typ     34 x       2      62 gates
nor02ii      tsmc035_typ      3 x       1       4 gates
oai21        tsmc035_typ      1 x       1       1 gates
xnor2        tsmc035_typ      4 x       2       8 gates

 Number of ports :                         44
 Number of nets :                          93
 Number of instances :                     68
 Number of references to this view :        1

Total accumulated area :
 Number of gates :                        182
 Number of accumulated instances :         68

*****************************************************
Cell:     sd_scan_driver_17_1_1110000000000000_1111111111111111
View: INTERFACE    Library: work
*****************************************************
 Cell        Library   References     Total Area
buf02        tsmc035_typ      3 x       1       3 gates
dffs_ni      tsmc035_typ     17 x       6      99 gates
inv01        tsmc035_typ      1 x       1       1 gates
inv02        tsmc035_typ      3 x       1       2 gates
latch        tsmc035_typ      1 x       2       2 gates
mux21_ni     tsmc035_typ     34 x       2      62 gates
nand02       tsmc035_typ      1 x       1       1 gates
nor02ii      tsmc035_typ      3 x       1       4 gates
xnor2        tsmc035_typ      2 x       2       4 gates
```

```
xor2         tsmc035_typ     3 x       2       6 gates

 Number of ports :                         40
 Number of nets :                          90
 Number of instances :                     68
 Number of references to this view :        1

Total accumulated area :
 Number of gates :                        184
 Number of accumulated instances :         68

*****************************************************
Cell: lc_lbist_control_3_0000_0001_0010_0011_0100_0101_0110_0111_
1000_1001_1010_1011_1    View: INTERFACE    Library: work
*****************************************************
 Cell        Library  References     Total Area
and02        tsmc035_typ      1 x       1       1 gates
and03        tsmc035_typ      1 x       2       2 gates
and04        tsmc035_typ      1 x       2       2 gates
aoi21        tsmc035_typ      2 x       1       2 gates
aoi22        tsmc035_typ      1 x       1       1 gates
aoi32        tsmc035_typ      2 x       2       3 gates
aoi322       tsmc035_typ      1 x       2       2 gates
aoi332       tsmc035_typ      1 x       2       2 gates
aoi43        tsmc035_typ      1 x       2       2 gates
buf02        tsmc035_typ      3 x       1       3 gates
dffr         tsmc035_typ     33 x       6     184 gates
dffs_ni      tsmc035_typ      2 x       6      12 gates
inv01        tsmc035_typ     18 x       1      14 gates
inv02        tsmc035_typ      1 x       1       1 gates
mux21        tsmc035_typ      2 x       2       4 gates
mux21_ni     tsmc035_typ     17 x       2      31 gates
nand02       tsmc035_typ     11 x       1      11 gates
nand02_2x    tsmc035_typ      1 x       1       1 gates
nand03       tsmc035_typ      4 x       1       5 gates
nand04       tsmc035_typ      2 x       1       3 gates
nor02_2x     tsmc035_typ      4 x       1       4 gates
nor02ii      tsmc035_typ      5 x       1       6 gates
nor03_2x     tsmc035_typ      6 x       1       7 gates
oai21        tsmc035_typ      8 x       1      10 gates
oai22        tsmc035_typ      1 x       1       1 gates
oai32        tsmc035_typ      3 x       2       5 gates
oai33        tsmc035_typ      1 x       2       2 gates
oai44        tsmc035_typ      1 x       3       3 gates
or02         tsmc035_typ      4 x       1       5 gates
```

```
or03          tsmc035_typ     1 x       2       2 gates
xnor2         tsmc035_typ     8 x       2      15 gates
xor2          tsmc035_typ     1 x       2       2 gates

 Number of ports :                        42
 Number of nets :                        177
 Number of instances :                   148
 Number of references to this view :       1

Total accumulated area :
 Number of gates :                       350
 Number of accumulated instances :       148

*******************************************************
Cell: co_clock_control_3_4_1    View: INTERFACE    Library: work
*******************************************************
 Cell       Library  References     Total Area
and02       tsmc035_typ      5 x     1       6 gates
and03       tsmc035_typ      2 x     2       3 gates
dffr        tsmc035_typ      9 x     6      50 gates
inv01       tsmc035_typ      3 x     1       2 gates
latch       tsmc035_typ      5 x     2      12 gates
nand03      tsmc035_typ      1 x     1       1 gates
nor02ii     tsmc035_typ      5 x     1       6 gates

 Number of ports :                        22
 Number of nets :                         41
 Number of instances :                    30
 Number of references to this view :       1

Total accumulated area :
 Number of gates :                        82
 Number of accumulated instances :        30

*******************************************************
Cell:  li_lbist_interface_3_17_17_17_000_4       View:  INTERFACE
Library: work
*******************************************************
 Cell        Library  References     Total Area
and02        tsmc035_typ      1 x     1       1 gates
and04        tsmc035_typ      8 x     2      14 gates
aoi22        tsmc035_typ     17 x     1      25 gates
dffr         tsmc035_typ     26 x     6     145 gates
dffs_ni      tsmc035_typ      3 x     6      17 gates
inv01        tsmc035_typ      1 x     1       1 gates
```

```
inv02        tsmc035_typ     8 x      1      6 gates
mux21        tsmc035_typ     3 x      2      5 gates
mux21_ni     tsmc035_typ     5 x      2      9 gates
nand02       tsmc035_typ    18 x      1     18 gates
nand03       tsmc035_typ     1 x      1      1 gates
nand04       tsmc035_typ     1 x      1      1 gates
nor02_2x     tsmc035_typ     4 x      1      4 gates
oai21        tsmc035_typ    22 x      1     27 gates
or02         tsmc035_typ     2 x      1      2 gates

 Number of ports :                       75
 Number of nets :                       152
 Number of instances :                  120
 Number of references to this view :      1

Total accumulated area :
 Number of gates :                      279
 Number of accumulated instances :      120

*******************************************************
Cell: bc_bist_controller    View: INTERFACE    Library: work
*******************************************************
 Cell                                                    Library
References     Total Area

and04                                                   tsmc035_
typ      1 x       2       2 gates
ao21                                                    tsmc035_
typ      1 x       2       2 gates
buf02                                                   tsmc035_
typ      8 x       1       8 gates
co_clock_control_3_4_1                                      work
1 x      82      82 gates
dffr                                                    tsmc035_
typ      5 x       6      28 gates
fake_gnd                                                tsmc035_
typ      1 x       1       1 fake_gnd
inv01                                                   tsmc035_
typ      3 x       1       2 gates
latch                                                   tsmc035_
typ      1 x       2       2 gates
lc_lbist_control_3_0000_0001_0010_0011_0100_0101_0110_0111_
1000_1001_1010_1011_1     work            1 x     350     350 gates
li_lbist_interface_3_17_17_17_000_4                         work
1 x     279     279 gates
```

```
mux21_ni                                                  tsmc035_
typ      7 x      2      13 gates
nand02                                                    tsmc035_
typ      2 x      1       2 gates
nand03                                                    tsmc035_
typ      1 x      1       1 gates
nor02_2x                                                  tsmc035_
typ      2 x      1       2 gates
nor02ii                                                   tsmc035_
typ      2 x      1       2 gates
nor03_2x                                                  tsmc035_
typ      2 x      1       2 gates
nor04                                                     tsmc035_
typ      3 x      1       4 gates
oai21                                                     tsmc035_
typ      2 x      1       2 gates
oai33                                                     tsmc035_
typ      1 x      2       2 gates
or02                                                      tsmc035_
typ      2 x      1       2 gates
or03                                                      tsmc035_
typ      2 x      2       3 gates
pc_pattern_count_2_4_4_15_000000000000000_111111111111111
work            1 x    381    381 gates
sd_scan_driver_17_1_11100000000000000_11111111111111111
work            1 x    184    184 gates
sm_scan_monitor_1_0_17_11100000000000000_11111111111111111
work            1 x    182    182 gates
xnor2                                                     tsmc035_
typ      1 x      2       2 gates
xor2                                                      tsmc035_
typ      1 x      2       2 gates

 Number of ports :                          31
 Number of nets :                          155
 Number of instances :                      54
 Number of references to this view :         1

Total accumulated area :
 Number of fake_gnd :                        1
 Number of gates :                        1542
 Number of accumulated instances :         657

*****************************************************
Cell: cdcscan    View: INTERFACE    Library: work
```

```
*****************************************************
 Cell        Library  References      Total Area
and02        tsmc035_typ      7 x      1        9 gates
inv01        tsmc035_typ      1 x      1        1 gates
inv02        tsmc035_typ      3 x      1        2 gates
latch        tsmc035_typ      1 x      2        2 gates
mux21        tsmc035_typ      1 x      2        2 gates
or02         tsmc035_typ      3 x      1        4 gates
sffr_ni      tsmc035_typ      2 x      7       14 gates

 Number of ports :                          15
 Number of nets :                           33
 Number of instances :                      18
 Number of references to this view :         1

Total accumulated area :
 Number of gates :                          34

*****************************************************
Cell: cdcscan_bist     View: INTERFACE    Library: work
*****************************************************
 Cell                  Library  References      Total Area
bc_bist_controller     work       1 x      1       1 fake_gnd
                                        1542    1542 gates
cdcscan                work       1 x     34      34 gates

 Number of ports :                          24
 Number of nets :                           34
 Number of instances :                       2
 Number of references to this view :         0

Total accumulated area :
 Number of fake_gnd :                        1
 Number of gates :                        1576
 Number of accumulated instances :         658
```

Memory BIST

Almost every ASIC and SOC chip uses memory. It is common to have multiple memory blocks within a single ASIC or SOC chip. These memories must be tested to ensure they are functional. Since each memory can range from a small RAM block such as 8 Kbit to large RAM blocks with Mbit size, testing each memory block will consume a lot of tester time, thus increasing testing cost.

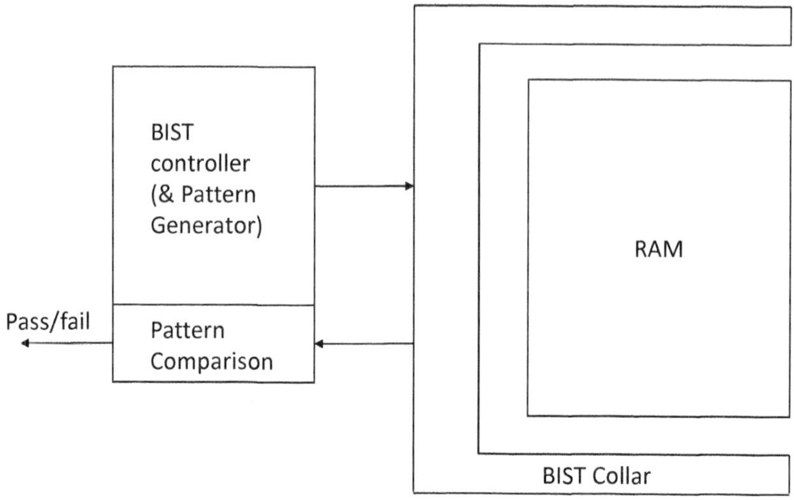

Fig. 5.27 Diagram showing block diagram for memory BIST

A common way to work around this problem is using memory BIST. In this method, a BIST controller can be used to pump in a certain pattern into the RAM block, followed by reading from the RAM block to ensure the pattern matches. This is repeated throughout the whole RAM. A read from the RAM is used to determine if the read and write matches. There are many algorithms available on the testing of memory BIST such as the popular March algorithm implemented in many memory BIST testings.

Figure 5.27 shows a block diagram for memory BIST. The memory RAM block is surrounded by a BIST collar. The BIST controller sends the main control signals to the BIST collar as well as the generated patterns based on the memory BIST testing algorithms. Examples of such algorithms are March, March2, March C-, and others. The patterns are written into the RAM block and then read out, and the pattern is compared to determine a pass or fail. The pass fail indicator can be an output pin to indicate to external systems that the memory BIST passes or fails.

For large design with multiple RAM blocks, multiple smaller memory RAM blocks that are physically close to each other can be combined together for testing using one MBIST controller, as shown in Fig. 5.28.

The RTL Verilog code for a design with a single 256 × 16 RAM block is shown in Example 5.7, and its top level block diagram generated by Mentor Graphics' HDL Designer shown in Fig. 5.29.

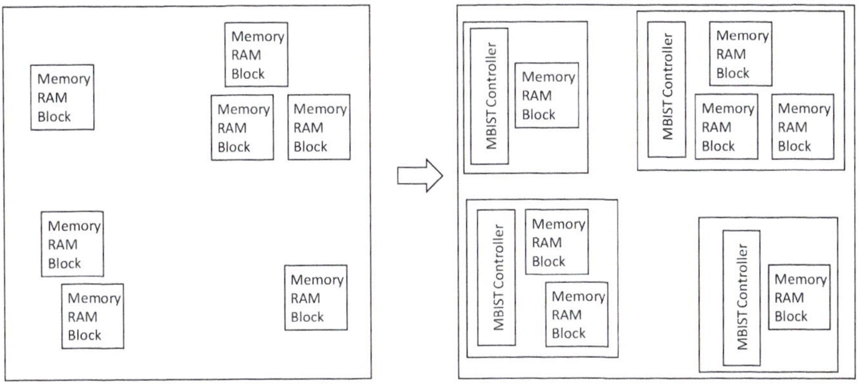

Fig. 5.28 Diagram showing memory BIST for design with multiple RAM blocks

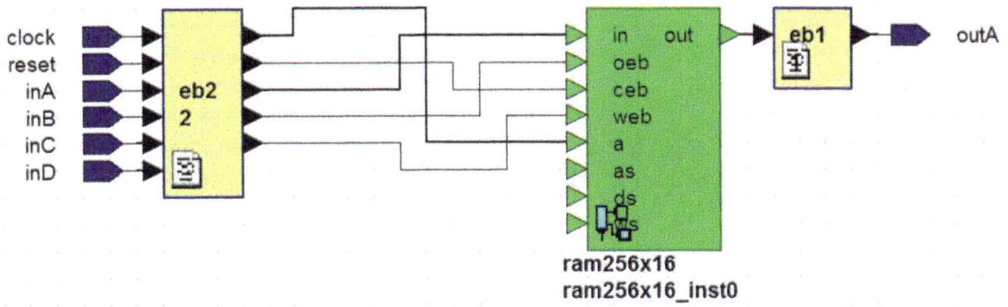

Fig. 5.29 Diagram showing top level block diagram view of Example 5.7

Example 5.7 Verilog code for a design using a 256 × 16 RAM block

```
module designwithram (clock, reset, inA, inB, inC, inD, outA);
input clock, reset;
input inA, inB;
input [7:0] inC;
input [15:0] inD;
output [15:0] outA;

reg ceb, oeb, web;
reg [7:0] address;
reg [15:0] data_in;
wire [15:0] data_out;

assign outA = data_out;

always @ (posedge clock or posedge reset)
```

```
begin
       if (reset)
              begin
                     ceb <= 1;
                     oeb <= 1;
                     web <= 1;
                     address <= 0;
                     data_in <= 0;
              end
       else
              begin
                     if (~inA & inB)
                            begin
                                   ceb <= 0;
                                   oeb <= 1;
                                   web <= 0;
                                   address <= inC;
                                   data_in <= inD;
                            end
                     else if (inA & ~inB)
                            begin
                                   ceb <= 0;
                                   web <= 1;
                                   oeb <= 0;
                                   address <= inC;
                                   data_in <= 0;
                            end
                     else
                            begin
                                   ceb <= 1;
                                   web <= 1;
                                   oeb <= 1;
                                   address <= 0;
                                   data_in <= 0;
                            end
              end
end

ram256x16 ram256x16_inst0 (
.a(address),
.as(),
.ceb(ceb),
.ds(),
.in(data_in),
.oeb(oeb),
```

```
.out(data_out),
.web(web),
.ws());

endmodule
```

Using the RTL Verilog of Example 5.7, the RTL code is put through Mentor Graphics MBISTArchitect tool to generate the memory BIST. Memory BIST can be inserted at RTL level or at gate level netlist. The output from the memory BIST tool such as MBISTArchitect is a modified RTL Verilog or gate level Verilog with RTL of the inserted memory BIST. The inserted memory BIST controller and its collar can be synthesized with the rest of the design logic through top-down synthesis method or can be individually synthesized using bottom-up methodology (Fig. 5.30).

The RTL Verilog of Example 5.7 is modified by MBISTArchitect to include the memory BIST controller as shown in Fig. 5.31. Additional signals are inserted into

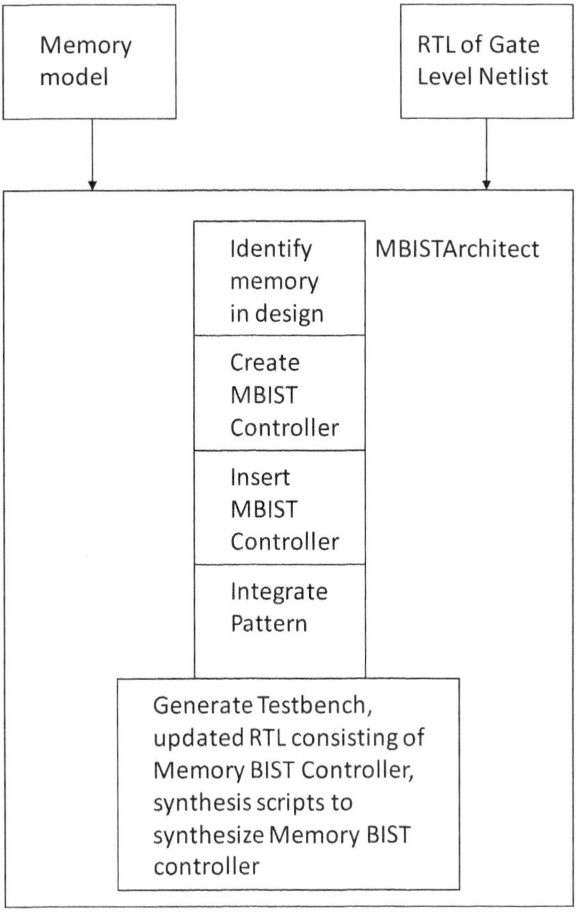

Fig. 5.30 Diagram showing flow to generate memory BIST controller of Example 5.7

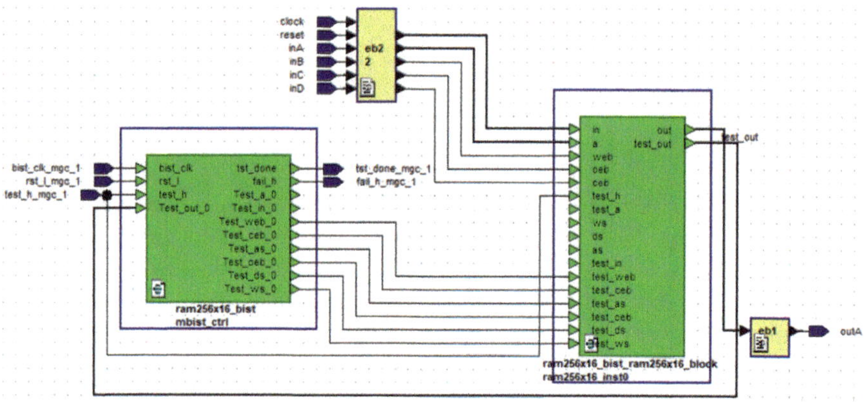

Fig. 5.31 Diagram showing top level block diagram view of Example 5.7 after memory BIST insertion

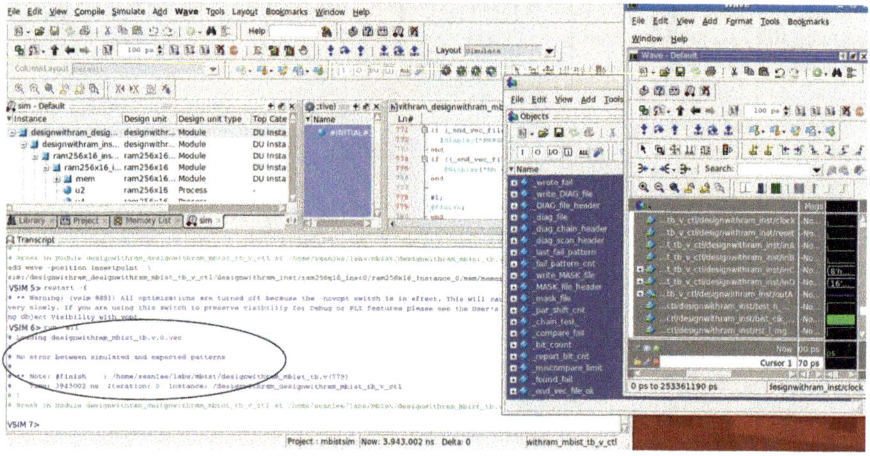

Fig. 5.32 Diagram showing simulation of testbench of Example 5.7 after memory BIST insertion

the RTL of Example 5.7 for the input and output pins required for memory BIST and insertion of memory BIST controller block.

MBISTArchitect also generates a Verilog testbench that can simulate the top level design module of Example 5.7 with the memory BIST inserted. Figures 5.32 and 5.33 show the simulation waveform of the generated memory BIST.

Referring to Fig. 5.32, the memory BIST writes all the addresses of the 256 × 16 RAM block and then reads it for matching. And when all the memory addresses have been checked, the simulation ends with the message displayed "No error

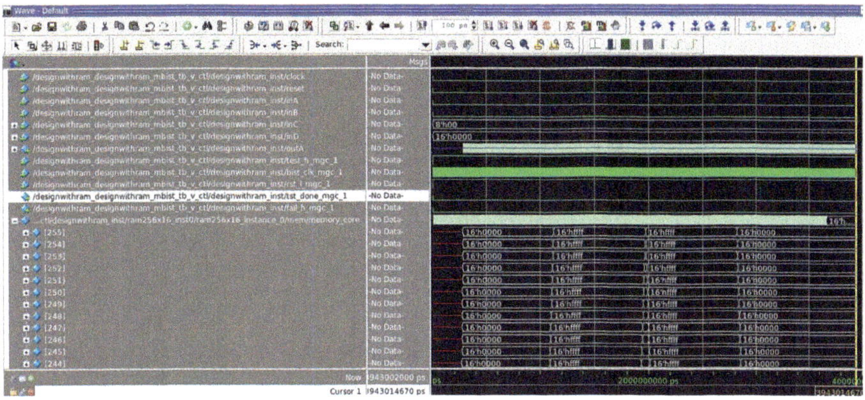

Fig. 5.33 Diagram showing end of simulation with no error between simulated and expected patterns

between simulated and expected patterns" as shown in Fig. 5.33. In actual usage of the memory BIST within an ASIC or SOC, the memory BIST controller will pump in the patterns into the memory addresses, reads it, and compares the patterns. An output pin from the ASIC or SOC is used as an indicator to the system level of a pass or fail condition.

Chapter 6
Signed Verilog

In RTL coding, when a wire or reg is declared for a signal, by default the signal is unsigned. If a signed representation of the wire or reg is needed, the Verilog keyword "signed" is used.

```
reg signed [15:0] b;
wire signed [7:0] d;
```

When a signal is defined as signed, the MSB of the signal determines if the value of the signal is positive or negative. A MSB of "1" represents a negative value, and a MSB of "0" represents a positive value as shown in Table 6.1.

Mixing Signed and Unsigned

When using signed keyword for RTL design, it is important to note all signals involved in the assignment must be signed. "outSum = inA + inB" with all three signals is being declared as signed. If either inA or inB is not declared as signed, outSum will not be signed even if it is declared as signed.

Mixing of signed and unsigned will result in unsigned. This is a common design mistake when writing RTL code for design that requires signed functionality. For example, when designing a module for complex control of brushless dc (BLDC) motor commutation, signed functionality is required to enable control of the motor commutation to differentiate a clockwise movement and counterclockwise movement of the motor. If any part of the RTL code is missing a signed keyword on a particular signal that requires signed functionality, the result of the logic function would be unsigned, thus causing a design bug that may be difficult to trace.

Example 6.1 shows the RTL code for a signed arithmetic module which does a signed addition, signed subtraction, and signed multiplication. Example 6.2 is a

W. F. Lee, *Learning from VLSI Design Experience*,
https://doi.org/10.1007/978-3-030-03238-8_6

Table 6.1 Table showing representation of signed numbers

Value	Signed binary [3:0]	Unsigned binary [3:0]
2	0010	0010
1	0001	0001
0	0000	0000
−1	1111	Not applicable
−2	1110	Not applicable

testbench with a for loop to pump in 256 combinations for inA, inB, and inC for the simulation.

Example 6.1 RTL code for signed arithmetic

```
module signed_arith (inA, inB, inC, outSum, outSub, outMul,
outunsigned);
input signed [3:0] inA, inB;
input [3:0] inC;
output signed [4:0] outSum, outSub;
output signed [7:0] outMul;
output signed [4:0] outunsigned;

wire signed [4:0] outSum, outSub;
wire signed [7:0] outMul;
wire signed [4:0] outunsigned;

assign outSum = inA + inB;
assign outSub = inA - inB;
assign outMul = inA * inB;
assign outunsigned = inA + inC;

endmodule
```

Example 6.2 Example showing testbench to simulate signed arithmetic

```
module singed_arith_tb ();

reg signed [3:0] inA, inB;
reg [3:0] inC;
wire signed [4:0] outSum, outSub;
wire signed [7:0] outMul;
wire signed [4:0] outunsigned;

integer i,j;
```

```
initial
begin
        for (i=0; i<16; i=i+1)
        begin
                for (j=0; j<16; j=j+1)
                begin
                        inA = i;
                        inB = j;
                        inC = j;
                        #10;
                end
        end
end

signed_arith signed_arith_inst (inA, inB, inC, outSum, outSub,
outMul, outunsigned);

endmodule
```

Referring to Example 6.1, an output signal outunsigned is declared signed and is the sum of inA and inC, whereby inA is also signed but inC is unsigned.

```
assign outunsigned = inA + inC;
```

When signed and unsigned is mixed, the result is by default unsigned. Therefore, although outunsigned is declared as signed, its result is unsigned.

Referring to the simulation results of testbench shown in Example 6.2, from the waveform of Fig. 6.1:

(a) When inA = 4’b0111, inB = 4’b1111, and outSum = 5’b00110

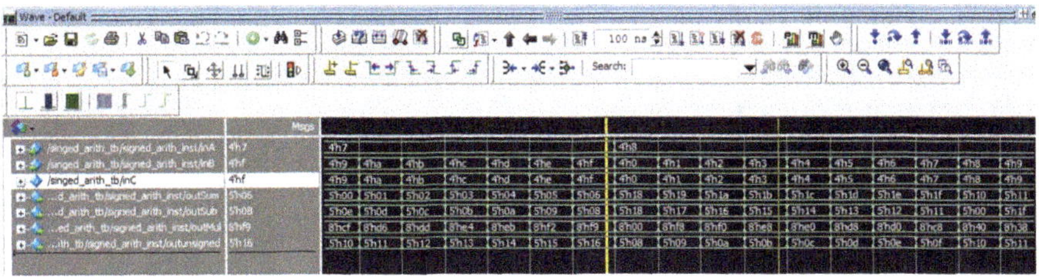

Fig. 6.1 Diagram showing simulation of signed Verilog and mixing of signed and unsigned Verilog

inA	2's complement inA	inB	2's complement inB	inA + inB
0111 (7)				0111 (7)
		1111	0001 (−1)	+1111 (−1)
				0110 (6)

(b)When inA = 4'b1000, inB = 4'b0010, and outSum = 5'b11010

inA	2's complement inA	inB	2's complement inB	inA + inB
1000	1000 (−8)			1000 (−8)
		0010 (2)		+0010 (2)
				11010 (−6)

(c) When inA = 4'b0111, inB = 4'b1101, and outSub = 5'b01010

inA	2's complement inA	inB	2's complement inB	inA − inB
0111 (7)				0111 (7)
		1101	0011 (−3)	−1101 (−3)
				01010 (10)

(d)When inA = 4'b1000, inB = 4'b0110, and outSub = 5'b10010

inA	2's complement inA	inB	2's complement inB	inA − inB
1000	1000 (−8)			1000 (−8)
		0110 (6)		−0110 (6)
				10010 (−14)

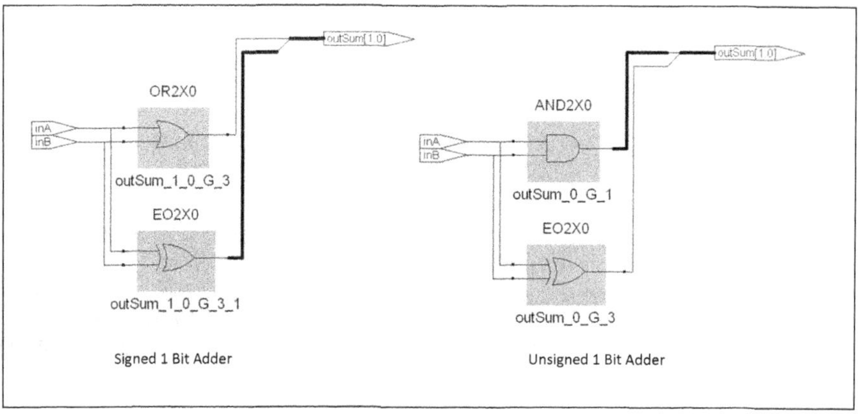

Fig. 6.2 Diagram showing schematic for 1 bit signed adder and 1 bit unsigned adder

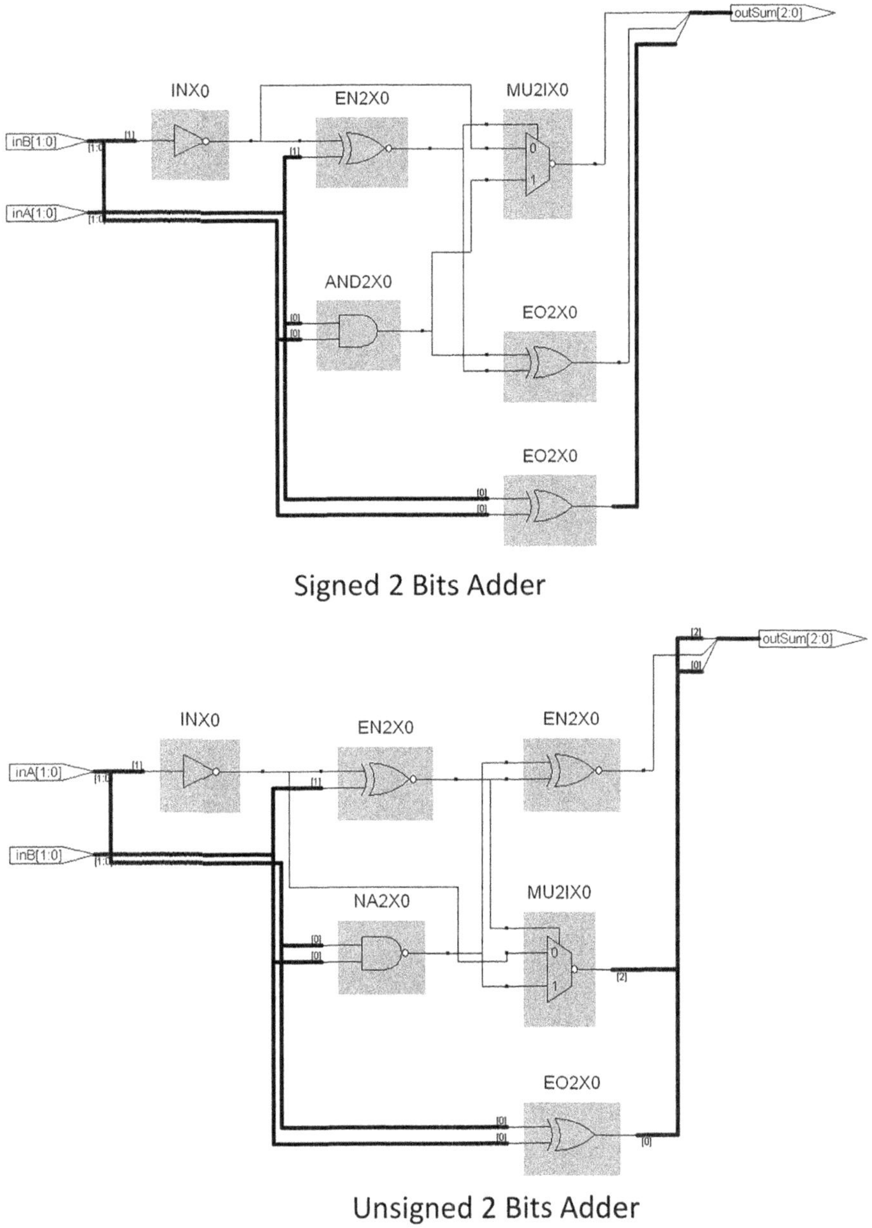

Fig. 6.3 Diagram showing schematic for 2 bit signed adder and 2 bit unsigned adder

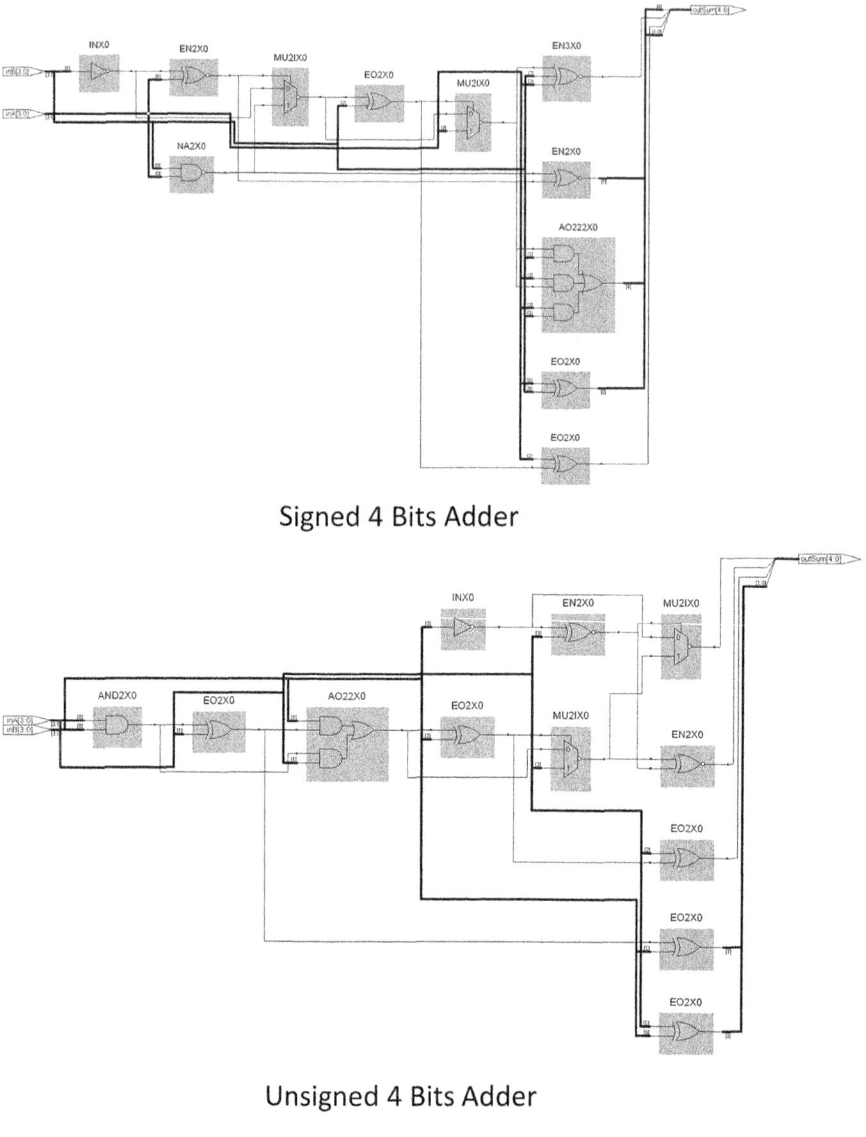

Fig. 6.4 Diagram showing schematic for 4 bit signed adder and 4 bit unsigned adder

(e) When inA = 4’b1000, inB = 4’b0101, and outMul = 8’b11011000

inA	2’s complement inA	inB	2’s complement inB	inA * inB
1000	1000 (−8)			1000 (−8)
		0101 (5)		*0101 (5)
				11011000 (−40)

(f) When inA = 4'b0111, inC = 4'b1010, and outunsigned = 5'b10001, InA and outunsigned are both declared as signed, but inC is unsigned. The functionality "outunsigned = inA + inC" is treated as unsigned when mixing signed and unsigned.

inA	2's complement inA	inC	2's complement inB	inA + inC
0111 (7)				0111 (7)
		1010 (10)	0110 (−6)	+1010 (10)
				10001 (17)

(g) When inA = 4'b1000, inC = 4'b1001, and outunsigned = 5'b10001, InA and outunsigned are both declared as signed, but inC is unsigned. The functionality "outunsigned = inA + inC" is treated as unsigned when mixing signed and unsigned.

inA	2's complement inA	inC	2's complement inC	inA + inC
1000 (8)	1000 (−8)			1000 (8)
		1001 (9)	0111 (−7)	+1001 (9)
				10001 (17)

It is important to note that mixing of signed and unsigned will result in the functionality being unsigned even though the functionality is declared as signed.

Figures 6.2, 6.3, and 6.4 show the schematic for a 1 bit signed adder and 1 bit unsigned adder, 2 bits and 4 bits, respectively. The amount of logic gates used to build a signed adder or unsigned adder is similar and not largely different.

Multiplication and Division of Signed and Unsigned Values

In logic arithmetic, addition and subtraction are commonly used for digital design. Subtraction is similar to addition except that the subtracted number is 2's complement. By using 2's complement for the subtracted number, both addition and subtraction can be unified to using addition only.

Multiplication and division are more complex and require more logic gates to be implemented in digital design. However if a multiplication is with 2^n, multiplication can be achieved by a simple shift left 1 bit for multiply 2 and shift left 2 bits for multiply 4. When shift left, the LSB is filled with 0, for example:

(a) Unsigned numbers (MSB filled with 0)

1.	4 * 2	= 8'b00000100 * 2
		= shift left 1 bit
		= 8'b00001000
		= 8
2.	6 * 4	= 8'b00000110 * 4
		= shift left 2 bits
		= 8'b00011000
		= 24

(b)Signed numbers (MSB sign extended)

1.	−6 * 2	= 8'b11111010 * 2
		= shift left 1 bit
		= 8'b11110100
		= −12
2.	−6 * 4	= 8'b11111010 * 4
		= shift left 2 bits
		= 8'b11101000
		= −24

Division with 2^n can be achieved by shift right 1 bit for divide 2 and shift right 2 bits for divide 4. When shift right, if the number is unsigned, the MSB is filled with 0. If the number is signed, MSB is sign extended, for example:

(c) Unsigned numbers (MSB filled with 0)

1.	4/2	= 8'b00000100/2
		= shift right 1 bit
		= 8'b00000010
		= 2
2.	6/4	= 8'b00000110/4
		= shift right 2 bits
		= 8'b00000001
		= 1

(d)Signed numbers (MSB sign extended)

1.	−12/2	= 8'b11110100/2
		= shift right 1 bit
		= 8'b11111010
		= −6
2.	−30/4	= 8'b11100010/4
		= shift right 2 bits
		= 8'b11111000
		= −8

Unsigned Shifting in Verilog

Shift right can be used for division with shift right n bits equivalent to division by 2^n. Shift left is used for multiplication with shift left n bits equivalent to multiplication by 2^n.

When shift left for multiplication by 2^n, the LSB is appended with 0. When shift right for division by 2^n, the MSB is appended with 0.

Shift is represented by ≫ for shift right and ≪ for shift left.

(a) Shift left for multiplication by 2^n

1.	7 * 2	= 8'b00000111 ≪ 1
		= 8'b00001110
		= 14
2.	12 * 8	= 8'b00001100 ≪ 3
		= 8'b01100000
		= 96

(b)Shift right for division by 2^n

1.	16/2	= 8'b00010000 ≫ 1
		= 8'b00001000
		= 8
2.	56/8	= 8'b00111000 ≫ 3
		= 8'b00000111
		= 7
3.	184/8	= 8'b10111000 ≫ 3
		= 8'b00010111
		= 23

Using shift left for multiplication by 2^n is an efficient design as it only involves shifting of bits and appending the LSB with zero. Similarly shift right for division by 2^n is an efficient design as it shifts bits for division and the MSB is appended with zero. Example 6.3 shows RTL Verilog utilizing a shift right and shift left operator. Example 6.4 shows a Verilog testbench to simulate the shift right and shift left operator.

Example 6.3 Example showing Verilog code utilizing shift right and shift left

```
module shiftright (inA, mulby2, outA);
input [7:0] inA;
input mulby2;
output [7:0] outA;
wire [7:0] outA;
assign outA = mulby2 ? (inA << 1) : (inA >> 1);
endmodule
```

Example 6.4 Example showing testbench Verilog code to simulate shift right and shift left operator

```
module shiftright_tb ();
reg [7:0] inA;
reg mulby2;
wire [7:0] outA;

initial
begin
       mulby2 = 0;
       inA = 8'hfa;
       #10;
       inA = 8'ha3;
       #10;
       inA = 8'h75;
       #10;
       mulby2 = 1;
       inA = 8'h34;
       #10;
       inA = 8'hb3;
       #10;
       inA = 8'h94;
       #10;
end

shiftright   shiftright_inst   (.inA(inA),   .mulby2(mulby2),   .
outA(outA));

endmodule
```

Figure 6.5 shows the simulation results of the shift right and shift left operator of Example 6.3 using the testbench shown in Example 6.4.

Referring to simulation shown in Fig. 6.5:

(a)	inA = 8'hfa	= 8'b11111010 ≫ 1
		= 8'b01111101
		= 8'h7d
(b)	inA = 8'ha3	= 8'b10100011 ≫ 1
		= 8'b01010001
		= 8'h51
(c)	inA = 8'h75	= 8'b01110101 ≫ 1
		= 8'b00111010
		= 8'h3a
(d)	inA = 8'h34	= 8'b00110100 ≪ 1
		= 8'b01101000
		= 8'h68
(e)	inA = 8'hb3	= 8'b10110011 ≪ 1
		= 8'b01100110
		= 8'h66
		1. When 179 is multiplied by 2, the result is 358
		2. Only 8 bits are used in the binary representation which represents 0–255
		3. 358 overflows the binary representation, thus moving it to (358-256=102) 102 (hexadecimal 66)
		4. When such overflow occurs, a larger amount of bits is required for the representation
		5. 8'hb3 * 2 = 16'b0000000010110011 ≪ 1 = 16'b0000000101100110 = 358
(f)	inA = 8'h94	= 8'b10010100 ≪ 1
		= 8'b00101000
		= 8'h28
		1. When 148 is multiplied by 2, the result is 296
		2. Only 8 bits are used in the binary representation which represents 0–255
		3. 296 overflows the binary representation, thus moving it to (296 – 256 = 40) 40 (hexadecimal 28)
		4. When such overflow occurs, a larger amount of bits is required for the representation
		5. 8'h94 * 2 = 16'b0000000010010100 ≪ 1 = 16'b0000000100101000 = 296

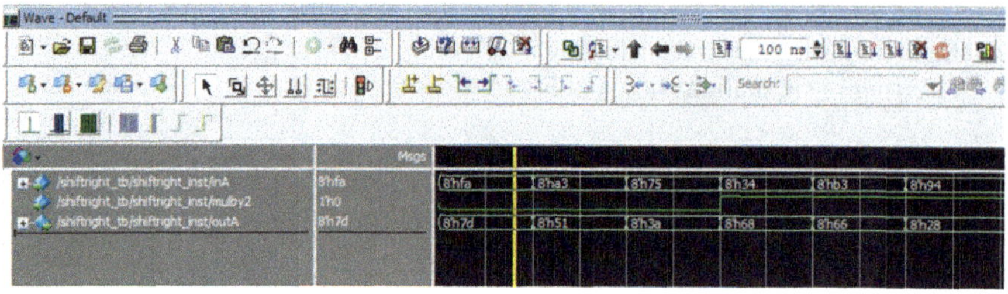

Fig. 6.5 Diagram showing simulation of shift right and shift left

Signed Shifting in Verilog

Signed shifting which consists of signed shift left and signed shift right behaves differently for signed numbers. Signed shift left behaves the same as unsigned shift left. However, signed shift right sign extends the MSB. Signed shift is represented by >>> for signed shift right and <<< for signed shift left.

(a)	−9 * 2	= 8'b11110111 ⋘ 1 = 8'b11101110 = −18
(b)	−24 * 4	= 8'b11101000 ⋘ 2 = 8'b10100000 = −96
(c)	−24 * 8	= 8'b11101000 ⋘ 3 = 8'b01000000 = 64
		1. When (−24) is multiplied by 8, the result is −192
		2. Only 8 bits are used in the signed binary representation which represents −128 to 127
		3. −192 overflows the binary representation, thus moving it to (−192 + 256 = 64) 64
		4. When such overflow occurs, a larger amount of bits is required for the representation
		5. −24 * 8 = 16'b1111111111101000 ⋘ 3 = 16'b1111111101000000 = −192
(d)	−100/4	= 8'b10011100 ⋙ 2 = 8'b11100111 = −25
(e)	−56/8	= 8'b11001000 ⋙ 3 = 8'b11111001 = −7

Example 6.5 shows RTL Verilog utilizing a signed shift right and signed shift left operator. Example 6.6 shows a Verilog testbench to simulate the signed shift right and signed shift left operator.

Example 6.5 Example showing Verilog code utilizing signed shift right and signed shift left

```
module signedshiftright (inA, mulby2, outA);
input signed [7:0] inA;
input mulby2;
output signed [7:0] outA;
wire signed [7:0] outA;
assign outA = mulby2 ? (inA <<< 1) : (inA >>> 1);
endmodule
```

Example 6.6 Example showing testbench Verilog code to simulate signed shift right and signed shift left operator

```
module signedshiftright_tb ();
reg signed [7:0] inA;
reg mulby2;
wire signed [7:0] outA;

initial
begin
       mulby2 = 0;
       inA = 8'hfa;
       #10;
       inA = 8'ha3;
       #10;
       inA = 8'h75;
       #10;
       mulby2 = 1;
       inA = 8'h34;
       #10;
       inA = 8'ha3;
       #10;
       inA = 8'h94;
       #10;
end

signedshiftright signedshiftright_inst (.inA(inA), .mulby2(mulby2),
.outA(outA));

endmodule
```

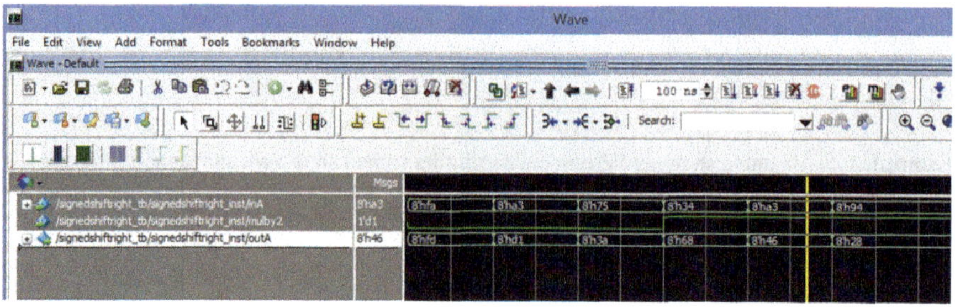

Fig. 6.6 Diagram showing simulation of signed shift right and signed shift left

Figure 6.6 shows the simulation results of the signed shift right and signed shift

(a)	inA = 8'hfa	= −6 ⋙ 1
		= 8'b11111010 ⋙ 1
		= 8'b11111101
		= 8'hfd
		= −3
(b)	inA = 8'ha3	= −93 ⋙ 1
		= 8'b10100011 ⋙ 1
		= 8'b11010001
		= 8'hd1
		= −47
		1. When (−93) is signed shift right (division), the result is −46.5 which is rounded down to −47
(c)	inA = 8'h75	= 8'b01110101 ⋙ 1
		= 8'b00111010
		= 8'h3a
(d)	inA = 8'h34	= 8'b00110100 ⋘ 1
		= 8'b01101000
		= 8'h68
(e)	inA = 8'ha3	= 8'b10100011 ⋘ 1
		= 8'b01000110
		= 8'h46
		1. When 8'ha3 (−93) is multiplied by 2, the result is −186
		2. Only 8 bits are used in the binary representation which represents 256; −186 overflows the binary representation, thus moving it to (−186 + 256 = 70) 70 (hexadecimal 46)
		3. When such overflow occurs, a larger amount of bits is required for the representation
		4. 8'ha3 * 2 = 16'b1111111110100011 ⋘ 1 = 16'b1111111101000110 = −186

(f)	inA = 8'h94	= 8'b10010100 ⋘ 1
		= 8'b00101000
		= 8'h28
		1. When 8'h94 (−108) is multiplied by 2, the result is −216
		2. Only 8 bits are used in the binary representation which represents 256; −216 overflows the binary representation, thus moving it to (−216 + 256 = 40) 40 (hexadecimal 28)
		3. When such overflow occurs, a larger amount of bits is required for the representation
		5. 8'h94 * 2 = 16'b1111111110010100 ⋘ 1 = 16'b1111111100101000 = −216

left operator of Example 6.5 using the testbench shown in Example 6.6.

Referring to simulation shown in Fig. 6.6:

Rounding Down Due to Signed Shift Right

Signed shift right ⋙ can be used for division by 2^n. However when a logic implementation of division by 2^n uses signed shift right, rounding down occurs as shown in Table 6.2.

For some logic implementation that uses signed shift right for continuous looping calculations such as fixed-point calculations, a rounding down is acceptable, but

Table 6.2 Table showing rounding down after signed shift right

Decimal	Binary								⋙1								Decimal after shift	Decimal/2
−57	1	1	0	0	0	1	1	1	1	1	1	0	0	0	1	1	−29	−57/2 = −28.5
−31	1	1	1	0	0	0	0	1	1	1	1	1	0	0	0	0	−16	−31/2 = −15.5
−15	1	1	1	1	0	0	0	1	1	1	1	1	1	0	0	0	−8	−15/2 = −7.5
−3	1	1	1	1	1	1	0	1	1	1	1	1	1	1	1	0	−2	−3/2 = −1.5
−1	1	1	1	1	1	1	1	1	1	1	1	1	1	1	1	1	−1	−1/2 = −0.5

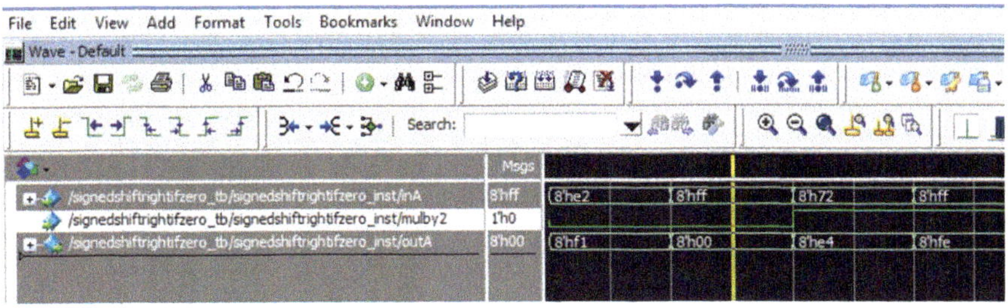

Fig. 6.7 Diagram showing simulation of signed shift right with if-else statement

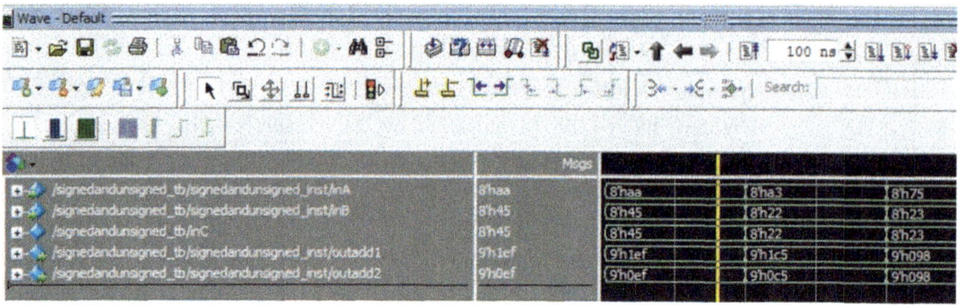

Fig. 6.8 Diagram showing simulation mixing of signed and unsigned in RTL Verilog

an error may occur when the calculations round down to −1 which is represented in Table 6.2 as 8'b11111111. A divide by 2 of −1 will yield the same result of −1 and will never reach 0. This creates an error which can accumulate over many calculation loops, thus causing functional failure. One such example is using signed shift right to implement calculations for a field-oriented control logic which involves looping to continuously update a signal through signed shift right for division. When the signal approaches −1 (8'b11111111), it will never reach 0 due to the nature of signed shift right of −1 yielding −1. A work-around for such condition is shown in the RTL code of Example 6.7.

Example 6.7 Example showing Verilog code using if-else in signed shift right

```
module signedshiftrightifzero (inA, mulby2, outA);
input signed [7:0] inA;
input mulby2;
output signed [7:0] outA;
reg signed [7:0] outA;

always @ (mulby2 or inA)
begin
        if (mulby2)
                outA = inA <<< 1;
        else // mulby2 = 0
                begin
                        if (inA == 8'hff)
                                outA = 0;
                        else
                                outA = inA >>> 1;
                end
end
endmodule
```

Example 6.8 Example showing testbench Verilog code to simulate if-else for signed shift right

```
module signedshiftrightifzero_tb ();
reg signed [7:0] inA;
reg mulby2;
wire signed [7:0] outA;

initial
begin
	mulby2 = 0;
	inA = 8'he2;
	#10;
	inA = 8'hff;
	#10;
	mulby2 = 1;
	inA = 8'h72;
	#10;
	inA = 8'hff;
	#10;
end

signedshiftrightifzero signedshiftrightifzero_inst (.inA(inA),
.mulby2(mulby2), .outA(outA));

endmodule
```

Figure 6.7 shows the simulation results for RTL Verilog code of Example 6.7 using the testbench shown in Example 6.8.

Referring to simulation shown in Fig. 6.7, when inA is 8'hff (−1 decimal), a signed shift right will result in 8'hff. The work-around shown in Example 6.7 uses the if-else statement to logically assign outA to 8'h00 when a signed shift right occurs on a 8'hff (−1 decimal).

Simulating RTL Using Signed and Unsigned

The RTL Verilog code in Example 6.3 shows unsigned shift, while RTL Verilog code in Examples 6.5–6.8 shows signed shift. When using keyword "signed" for RTL, it is important to remember a common mistake occurs when mixing signed and unsigned. Forgetting to use the signed keyword can mess up expected functionality.

Example 6.9 shows a RTL Verilog code with two adder outputs *outadd1* and *outadd2*. The output signal *outadd1* is a signed addition of signed *inA* and signed *inB*. However the output signal *outadd2* is a signed addition of signed *inA* and unsigned *inC*. Example 6.10 shows a Verilog testbench to simulate the RTL Verilog code mixing usage of signed and unsigned.

Example 6.9 Example showing Verilog code mixing usage of signed and unsigned

```
module signedandunsigned (inA, inB, inC, outadd1, outadd2);
input signed [7:0] inA, inB;
input [7:0] inC;
output signed [8:0] outadd1, outadd2;
wire signed [8:0] outadd1, outadd2;

assign outadd1 = inA + inB;
assign outadd2 = inA + inC;

endmodule
```

Example 6.10 Example showing Verilog testbench to simulate RTL code mixing usage of signed and unsigned

```
module signedandunsigned_tb ();
reg signed [7:0] inA, inB, inC;
wire signed [8:0] outadd1, outadd2;

initial
begin
       inA = 8'haa;
       inB = 8'h45;
       inC = 8'h45;
       #10;
       inA = 8'ha3;
       inB = 8'h22;
       inC = 8'h22;
       #10;
       inA = 8'h75;
       inB = 8'h23;
       inC = 8'h23;
       #10;
end

signedandunsigned signedandunsigned_inst (.inA(inA), .inB(inB),
.inC(inC), .outadd1(outadd1), .outadd2(outadd2));

endmodule
```

Referring to Example 6.10, inB and inC are pumped with the same stimulus. However inB is declared as signed in the RTL Verilog, while inC is declared as unsigned.

Referring to the simulation waveform shown in Fig. 6.8:

(a)	inA = 8'haa (signed), inB = 8'h45 (signed), and inC = 8'h45 (unsigned)
	1. outadd1 (signed) = 8'haa (−86 decimal) + 8'h45 (69 decimal) = 9'h1ef (−17 decimal)
	2. outadd2 (signed) = 8'haa (170 unsigned decimal) + 8'h45 (69 decimal) = 9'h0ef (239 decimal)
	3. outadd2 although defined as signed, its assignment of inA which is declared as signed with inC which is declared as unsigned results in the assignment being treated as unsigned
(b)	inA = 8'ha3 (signed), inB = 8'h22 (signed), inC = 8'h22 (unsigned)
	1. outadd1 (signed) = 8'ha3 (−93 decimal) + 8'h22 (34 decimal) = 9'h1c5 (−59 decimal)
	2. outadd2 (signed) = 8'ha3 (163 unsigned decimal) + 8'h22 (34 decimal) = 9'h0c5 (197 decimal)
	3. outadd2 is unsigned addition of inA with inC due to mixing of signed and unsigned resulting in the assignment treated as unsigned
(c)	inA = 8'h75 (signed), inB = 8'h23 (signed), inC = 8'h23 (unsigned)
	1. outadd1 (signed) = 8'h75 (117 decimal) + 8'h23 (35 decimal) = 9'h098 (152 decimal)
	2. outadd2 (signed) = 8'h75 (117 decimal) + 8'h23 (35 decimal) = 9'h098 (152 decimal)
	3. outadd2 is unsigned addition of inA with inC due to mixing of signed and unsigned resulting in the assignment treated as unsigned
	4. As inA, inB, and inC are positive numbers, outadd1 and outadd2 have the same addition value of 9'h098

When writing RTL Verilog to represent functionality of a design, it is important to ensure the signals involved in a signed assignment must be declared as signed. Otherwise the assignment is treated as unsigned. This may cause simulation failures between simulated waveforms and expected RTL functionality.

Chapter 7
State Machine

A design functionality which is process based, flow based, or state based can be designed using a state machine. A state machine is a synchronous design with a clock and a reset. The state machine moves from one state to another state based upon a clock cycle and predefined valid signals. The state machine can stay within the same state for a prolonged number of cycles if the predefined valid signal is not present and will only move to another state depending on the validity of the predefined signals.

A state machine can be classified as a one-hot state machine, a gray encoding state machine, or a binary encoding state machine. Whatever the encoding used to represent a state machine, the design of the state machine is similar with the exception of the encoding (Table 7.1).

RTL Verilog for a State Machine

There are three different ways to write the RTL Verilog for a state machine. Although these three different ways of RTL code have slight differences, the synthesized results are similar and synthesized to similar circuit.

Figure 7.1 shows the block diagram for a design with three validfier signals and a 4 bit output smoutput [3:0].

Table 7.2 shows the specification for the state transitions for the state machine.

Using the state transition specification of the design shown in Table 7.2, a state diagram is drawn to represent the validfier signals and how they cause the transition from one state to another state, as shown in Fig. 7.2 (state diagram captured from Mentor Graphics' HDL Designer). The RTL Verilog code is written based on the state transitions in the state diagram.

W. F. Lee, *Learning from VLSI Design Experience*,
https://doi.org/10.1007/978-3-030-03238-8_7

Table 7.1 Table showing the different encoding for a state machine

State	Binary encoding	Gray encoding	One-hot encoding
IDLE	000	000	00000001
STATEA	001	001	00000010
STATEB	010	011	00000100
STATEC	011	010	00001000
STATED	100	110	00010000
STATEE	101	111	00100000
STATEF	110	101	01000000
STATEG	111	100	10000000

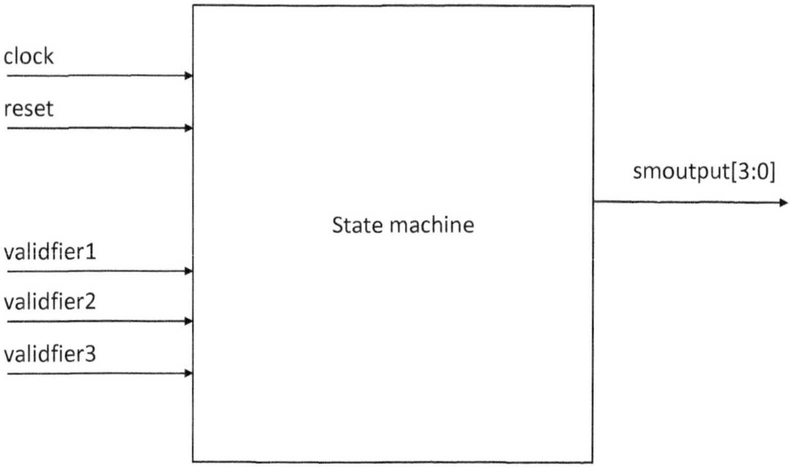

Fig. 7.1 Block diagram of state machine with validfier input signals

Example 7.1 Example showing Verilog code for state machine based on state diagram of Fig. 7.2

```
module statemachineone (clock, reset, validfier1, validfier2, valid-
fier3, smoutput);
input clock, reset;
input validfier1, validfier2, validfier3;
output [3:0] smoutput;
wire [3:0] smoutput;
reg [3:0] present_state, next_state;
parameter [3:0] IDLE=3'b000,
                STATE1=3'b001,
                STATE2=3'b010,
                STATE3=3'b011,
                STATE4=3'b100,
                STATE5=3'b101;
assign smoutput = present_state;
always @ (posedge clock or posedge reset)
```

Table 7.2 Table showing tate transition specification of the design

present_state	next_state	validfier1	validfier2	validfier3	smoutput [3:0]
IDLE	STATE1	1	1	X	000
IDLE	STATE3	1	X	1	000
IDLE	STATE4	1	1	1	000
STATE1	STATE5	X	1	1	001
STATE2	STATE1	1	0	X	010
STATE2	STATE4	X	1	0	010
STATE3	STATE2	0	X	0	011
STATE3	STATE4	X	1	X	011
STATE4	STATE5	X	0	X	100
STATE5	STATE3	0	X	0	101

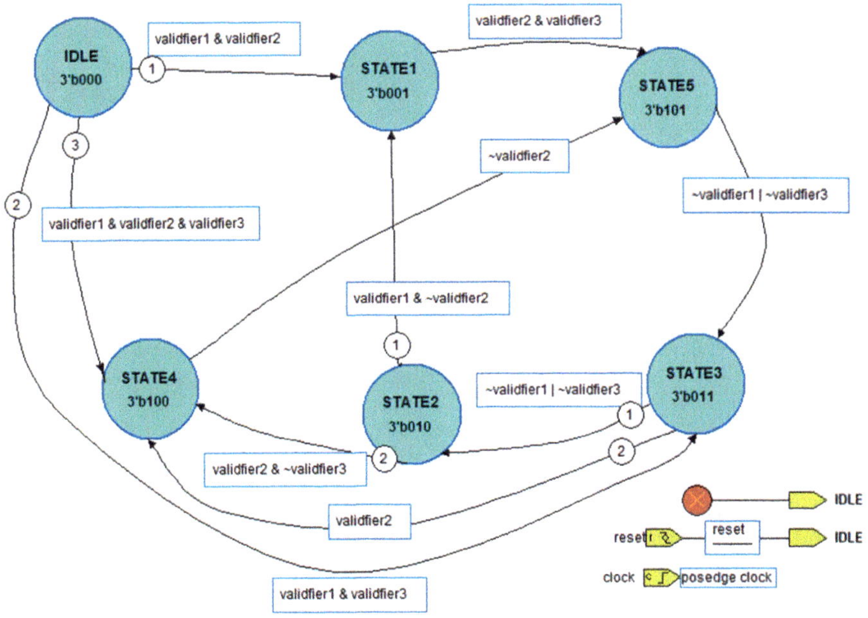

Fig. 7.2 State diagram for the state machine design

```
begin
        if (reset)
                present_state <= IDLE;
        else
                present_state <= next_state;
end

always @ (present_state or validfier1 or validfier2 or validfier3)
begin
        case (present_state)
                IDLE:
```

```
                    begin
                            if (validfier1 & validfier2)
                                    next_state = STATE1;
                            else if (validfier1 & validfier3)
                                    next_state = STATE3;
                             else if (validfier1 & validfier2 &
validfier3)
                                    next_state = STATE4;
                            else
                                    next_state = IDLE;
                    end
            STATE1:
                    begin
                            if (validfier2 & validfier3)
                                    next_state = STATE5;
                            else
                                    next_state = STATE1;
                    end
            STATE2:
                    begin ·
                            if (validfier1 & ~validfier2)
                                    next_state = STATE1;
                            else if (validfier2 & ~validfier3)
                                    next_state = STATE4;
                            else
                                    next_state = STATE2;
                    end
            STATE3:
                    begin
                            if (~validfier1 | ~validfier3)
                                    next_state = STATE2;
                            else if (validfier2)
                                    next_state = STATE4;
                            else
                                    next_state = STATE3;
                    end
            STATE4:
                    begin
                            if (~validfier2)
                                    next_state = STATE5;
                            else
                                    next_state = STATE4;
                    end
            STATE5:
                    begin
```

```
                        if (~validfier1 | ~validfier3)
                                next_state = STATE3;
                        else
                                next_state = STATE5;
                end
        default:
                begin
                        next_state = IDLE;
                end
    endcase
end
endmodule
```

Example 7.1 shows the RTL Verilog code for the state machine using binary encoding. If gray encoding is needed, the part of the RTL Verilog which defines the parameter for the states can be modified to gray encoding.

```
parameter [3:0] IDLE=3'b000,
                STATE1=3'b001,
                STATE2=3'b011,
                STATE3=3'b010,
                STATE4=3'b110,
                STATE5=3'b111;
```

For one-hot encoding, the 4 bit parameter must be expanded to 6 bits.

```
parameter [5:0] IDLE=6'b000001,
                STATE1=6'b000010,
                STATE2=6'b000100,
                STATE3=6'b001000,
                STATE4=6'b010000,
                STATE5=6'b100000;
```

One-hot encoding is the fastest state machine as the state decoding is only single bit to determine between one state and another state. However one-hot encoding is bigger as more bits are needed to represent the states. Therefore one-hot encoding needs more flip-flops to represent the state flops. A large state machine with 30 states will need 30 bits for one-hot state machine, which will synthesize to 30 state flip-flops. A binary encoding state machine with 30 states only needs 5 bits.

Therefore binary encoding and gray encoding for state machine result in smaller state machine as fewer state flip-flops are needed, but the state machine is slower. One-hot state machine is faster but is larger due to more state flip-flops needed.

Example 7.2 shows a Verilog testbench for simulating the state machine of Example 7.1.

Example 7.2 Example showing Verilog code of testbench for simulation of state machine RTL Verilog

```
module statemachineone_tb ();

reg clock, reset, validfier1, validfier2, validfier3;
wire [3:0] smoutput;

initial
begin
        clock = 0;
        forever clock = #10 !clock;
end

initial
begin
        reset = 0;
        validfier1 = 0;
        validfier2 = 0;
        validfier3 = 0;
        #10;
        reset = 1;
        #20;
        reset = 0;
        #10;
        // IDLE -> STATE1
        validfier1 = 1;
        validfier2 = 1;
        #20;
        // STATE1 -> STATE5
        validfier2 = 1;
        validfier3 = 1;
        #20;
        // STATE5 -> STATE3
        validfier1 = 0;
        validfier3 = 0;
        #20;
        // STATE3 -> STATE4
        validfier2 = 1;
        #20;
        // STATE4 -> STATE5
        validfier2 = 0;
        #20;
        // STATE5 -> IDLE;
        reset = 1;
```

```
        validfier1 = 0;
        validfier2 = 0;
        validfier3 = 0;
        #20;
        reset = 0;
        #20;
        // IDLE -> STATE3;
        validfier1 = 1;
        validfier3 = 1;
        #20;
        // STATE3 -> STATE4
        validfier2 = 1;
        #20;
        // STATE4 -> IDLE
        reset = 1;
        validfier1 = 0;
        validfier2 = 0;
        validfier3 = 0;
        #20;
        reset = 0;
        #20;
        // IDLE -> STATE4
        validfier1 = 1;
        validfier2 = 1;
        validfier3 = 1;
        #20;
end

statemachineone      statemachineone_inst      (.clock(clock),      .
reset(reset),   .validfier1(validfier1),   .validfier2(validfier2),   .
validfier3(validfier3), .smoutput(smoutput));

endmodule
```

The testbench in Example 7.2 simulates the RTL Verilog code of the state machine in Example 7.1. The simulation waveform is shown in Fig. 7.3. Referring to the waveform of Fig. 7.3:

(a) From reset, at rising edge of clock after reset goes low, and validfier1 and validfier2 are at logic 1, present_STATE (in the waveform of Fig. 7.3 is listed as convertedps to convert the STATE representation from binary values to state values) changes from IDLE to STATE1.
(b) On the next rising edge of clock, when validfier2 and validfier3 are at logic 1, thus the STATE machine changes from STATE1 to STATE5.
(c) On the next rising edge when present_STATE is at STATE5, and validfier1 or validfier3 is at logic 0, the state machine changes from STATE5 to STATE3.

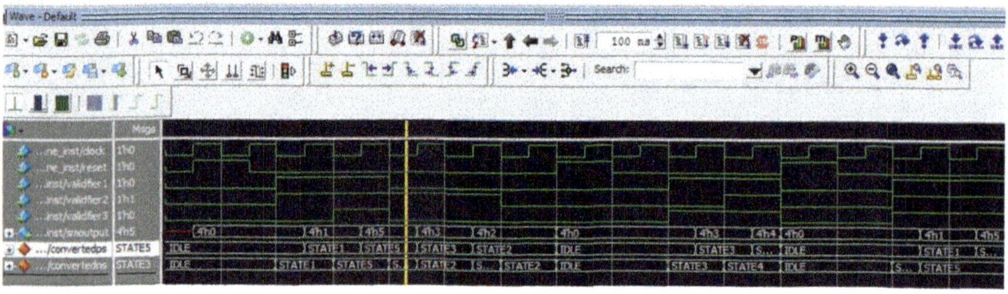

Fig. 7.3 Diagram showing simulation results of state machine for Example 7.1

(d) Each transition of the state machine RTL code is verified using the testbench shown in example 7.2. If any of the transitions are not exercised by the state machine and not verified, there could be a bug in the state machine transition. It is important that each state transition of a state machine is simulated and verified.

RTL Coding Style for State Machine Using Two Always Processes

Example 7.1 shows a commonly used coding style for writing the RTL for a state machine using two separate always processes. The first always process is responsible for the present_state assignment which translates to state flip-flops during synthesis. The second always process is responsible for the combinational logic to generate the next_state.

The combinational logic to generate the output of the state machine smoutput is assigned directly from present_state for Example 7.1. Some RTL designers may use a third always process to represent the combinational logic to generate smoutput.

Figure 7.4 shows a top-level schematic diagram of the logic that will be synthesized by the RTL Verilog code of Example 7.1. The output of the flip-flop is the signal present_state with the input to the flip-flop being next_state.

Different RTL Coding Styles for State Machine

There are three different methods of coding style to write the RTL for a state machine. Example 7.1 shows the RTL coding style using two always processes to represent the state flip-flops for generation of present_state and another always process to represent the combinational logic to generate next_state. Some RTL designers may use a third always process to generate the combinational logic of the state machine smoutput.

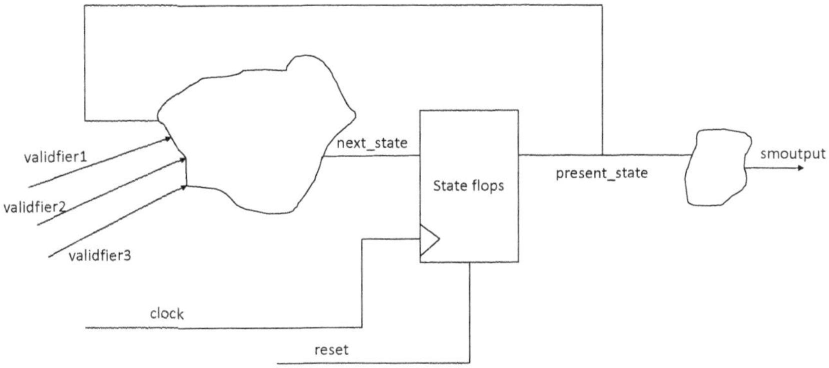

Fig. 7.4 Diagram showing schematics for a state machine showing present_state and next_state

A second method of coding style to write the RTL for the state machine uses one always process to generate a state signal. For this second method of coding style, the present_state and next_state are combined into one single signal state. Example 7.3 shows the RTL Verilog code for the state machine of Example 7.1, with same functionality but with a different coding style.

Example 7.3 Example showing RTL Verilog for state machine using one always process for state generation

```
module statemachinetwo (clock, reset, validfier1, validfier2, valid-
fier3, smoutput);
input clock, reset;
input validfier1, validfier2, validfier3;
output [3:0] smoutput;
wire [3:0] smoutput;
reg [3:0] state;
parameter [3:0] IDLE=3'b000,
                STATE1=3'b001,
                STATE2=3'b010,
                STATE3=3'b011,
                STATE4=3'b100,
                STATE5=3'b101;

assign smoutput = state;

always @ (posedge clock or posedge reset)
begin
        if (reset)
                state <= IDLE;
        else
        begin
```

```
case (state)
        IDLE:
                begin
                if (validfier1 & validfier2)
                        state <= STATE1;
                else if (validfier1 & validfier3)
                        state <= STATE3;
                else if (validfier1 & validfier2 &
                 validfier3)
                        state <= STATE4;
                else
                        state <= IDLE;
                end
        STATE1:
                begin
                if (validfier2 & validfier3)
                        state <= STATE5;
                else
                        state <= STATE1;
                end
        STATE2:
                begin
                if (validfier1 & ~validfier2)
                        state <= STATE1;
                else if (validfier2 & ~validfier3)
                        state <= STATE4;
                else
                        state <= STATE2;
                end
        STATE3:
                begin
                if (~validfier1 | ~validfier3)
                        state <= STATE2;
                else if (validfier2)
                        state <= STATE4;
                else
                        state <= STATE3;
                end
        STATE4:
                begin
                if (~validfier2)
                        state <= STATE5;
                else
                        state <= STATE4;
                end
```

```
                    STATE5:
                            begin
                            if (~validfier1 | ~validfier3)
                                    state <= STATE3;
                            else
                                    state <= STATE5;
                            end
                    default:
                            begin
                            state <= IDLE;
                            end
            endcase
        end
end
endmodule
```

The Verilog testbench shown in Example 7.4 is similar to the Verilog testbench of Example 7.2 which pumps in a similar stimulus but with the RTL instantiation being the RTL Verilog code of Example 7.3.

Example 7.4 Example showing Verilog code of testbench for simulation of state machine RTL Verilog using one always process for state generation

```
module statemachinetwo_tb ();

reg clock, reset, validfier1, validfier2, validfier3;
wire [3:0] smoutput;

initial
begin
        clock = 0;
        forever clock = #10 !clock;
end

initial
begin
        reset = 0;
        validfier1 = 0;
        validfier2 = 0;
        validfier3 = 0;
        #10;
        reset = 1;
        #20;
        reset = 0;
        #10;
```

```
// IDLE -> STATE1
validfier1 = 1;
validfier2 = 1;
#20;
// STATE1 -> STATE5
validfier2 = 1;
validfier3 = 1;
#20;
// STATE5 -> STATE3
validfier1 = 0;
validfier3 = 0;
#20;
// STATE3 -> STATE4
validfier2 = 1;
#20;
// STATE4 -> STATE5
validfier2 = 0;
#20;
// STATE5 -> IDLE;
reset = 1;
validfier1 = 0;
validfier2 = 0;
validfier3 = 0;
#20;
reset = 0;
#20;
// IDLE -> STATE3;
validfier1 = 1;
validfier3 = 1;
#20;
// STATE3 -> STATE4
validfier2 = 1;
#20;
// STATE4 -> IDLE
reset = 1;
validfier1 = 0;
validfier2 = 0;
validfier3 = 0;
#20;
reset = 0;
#20;
// IDLE -> STATE4
validfier1 = 1;
validfier2 = 1;
validfier3 = 1;
```

```
        #20;
end

statemachinetwo    statemachinetwo_inst    (.clock(clock),    .
reset(reset),   .validfier1(validfier1),   .validfier2(validfier2),   .
validfier3(validfier3), .smoutput(smoutput));

endmodule
```

Figure 7.5 shows the simulation waveform of testbench of Example 7.4, with both simulation waveforms of Figs. 7.3 and 7.5 showing the same functionality.

Figure 7.6 shows the top-level schematic diagram of the logic that will be synthesized by the RTL Verilog code of Example 7.3. The output of the flip-flop is the signal state. The schematic diagrams of Figs. 7.4 and 7.6 are similar with only minor naming changes on the signals.

The schematic shown in Fig. 7.6 does not have signals present_state or next_state as the RTL code of Example 7.3 uses only a single state signal to represent the state flop output. The input to the state flops is an arbitrary signal name given during synthesis.

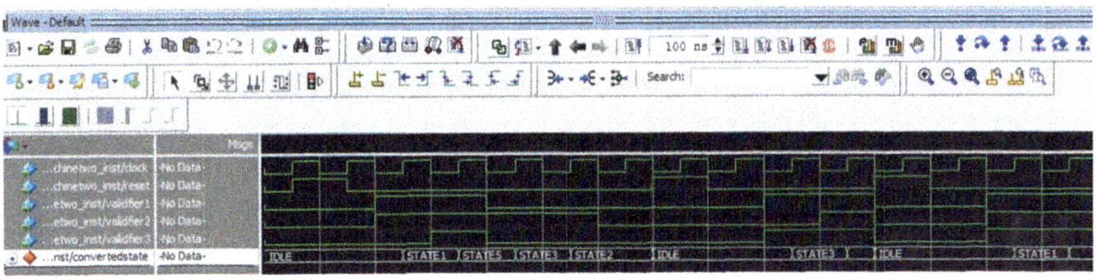

Fig. 7.5 Diagram showing simulation results of state machine for Example 7.4

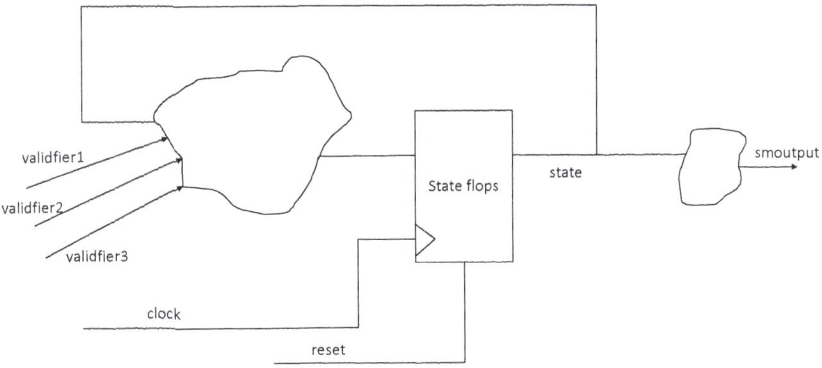

Fig. 7.6 Diagram showing schematics for a state machine of Example 7.3 showing signal state

A third method of coding style to write the RTL for the state machine uses function syntax. Although this method is not commonly used, it synthesizes to similar logic as the other two methods of RTL coding for state machine shown in Examples 7.1 and 7.3. For this third method of coding style, the present_state is used in a sequential statement for the state flops, while the next_state is generated from a function which defines the combinational logic for the state machine.

Example 7.5 shows the RTL Verilog code for the state machine using function with same functionality but with a different coding style.

Example 7.5 Example showing RTL Verilog for state machine using function for next_state generation

```
module statemachinethree (clock, reset, validfier1, validfier2,
validfier3, smoutput);
input clock, reset;
input validfier1, validfier2, validfier3;
output [3:0] smoutput;
wire [3:0] smoutput;
reg [3:0] present_state;
wire [3:0] next_state;
parameter [3:0] IDLE=3'b000,
                STATE1=3'b001,
                STATE2=3'b010,
                STATE3=3'b011,
                STATE4=3'b100,
                STATE5=3'b101;

assign smoutput = present_state;
assign next_state = ns (reset, present_state, validfier1, valid-
fier2, validfier3);

always @ (posedge clock or posedge reset)
begin
        if (reset)
                present_state <= IDLE;
        else
                present_state <= next_state;
end

function [3:0] ns;
input reset;
input [3:0] present_state;
input validfier1, validfier2, validfier3;
begin
        if (reset)
```

```
            ns = IDLE;
      else
      begin
            case (present_state)
                  IDLE:
                        begin
                        if (validfier1 & validfier2)
                              ns = STATE1;
                        else if (validfier1 & validfier3)
                              ns = STATE3;
                        else if (validfier1 & validfier2 &
                        validfier3)
                              ns = STATE4;
                        else
                              ns = IDLE;
                        end
                  STATE1:
                        begin
                        if (validfier2 & validfier3)
                              ns = STATE5;
                        else
                              ns = STATE1;
                        end
                  STATE2:
                        begin
                        if (validfier1 & ~validfier2)
                              ns = STATE1;
                        else if (validfier2 & ~validfier3)
                              ns = STATE4;
                        else
                              ns = STATE2;
                        end
                  STATE3:
                        begin
                        if (~validfier1 | ~validfier3)
                              ns = STATE2;
                        else if (validfier2)
                              ns = STATE4;
                        else
                              ns = STATE3;
                        end
                  STATE4:
                        begin
                        if (~validfier2)
                              ns = STATE5;
```

```
                                else
                                        ns = STATE4;
                                end
                        STATE5:
                                begin
                                if (~validfier1 | ~validfier3)
                                        ns = STATE3;
                                else
                                        ns = STATE5;
                                end
                        default:
                                begin
                                ns = IDLE;
                                end
                endcase
        end
end
endfunction
endmodule
```

The Verilog testbench shown in Example 7.6 is similar to the Verilog testbench of Examples 7.2 and 7.4 which pump in a similar stimulus but with the RTL instantiation being the RTL Verilog code of Example 7.5.

Example 7.6 Example showing Verilog code of testbench for simulation of state machine RTL Verilog using function for next_state generation

```
module statemachinethree_tb ();

reg clock, reset, validfier1, validfier2, validfier3;
wire [3:0] smoutput;

initial
begin
        clock = 0;
        forever clock = #10 !clock;
end

initial
begin
        reset = 0;
        validfier1 = 0;
        validfier2 = 0;
        validfier3 = 0;
        #10;
```

```
reset = 1;
#20;
reset = 0;
#10;
// IDLE -> STATE1
validfier1 = 1;
validfier2 = 1;
#20;
// STATE1 -> STATE5
validfier2 = 1;
validfier3 = 1;
#20;
// STATE5 -> STATE3
validfier1 = 0;
validfier3 = 0;
#20;
// STATE3 -> STATE4
validfier2 = 1;
#20;
// STATE4 -> STATE5
validfier2 = 0;
#20;
// STATE5 -> IDLE;
reset = 1;
validfier1 = 0;
validfier2 = 0;
validfier3 = 0;
#20;
reset = 0;
#20;
// IDLE -> STATE3;
validfier1 = 1;
validfier3 = 1;
#20;
// STATE3 -> STATE4
validfier2 = 1;
#20;
// STATE4 -> IDLE
reset = 1;
validfier1 = 0;
validfier2 = 0;
validfier3 = 0;
#20;
reset = 0;
#20;
```

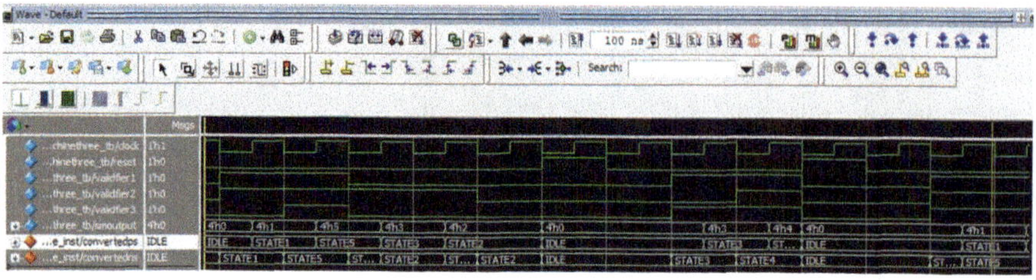

Fig. 7.7 Diagram showing simulation results of state machine for Example 7.6

```
        // IDLE -> STATE4
        validfier1 = 1;
        validfier2 = 1;
        validfier3 = 1;
        #20;
end

statemachinethree    statemachinethree_inst    (.clock(clock),    .
reset(reset),    .validfier1(validfier1),    .validfier2(validfier2),    .
validfier3(validfier3), .smoutput(smoutput));

endmodule
```

Figure 7.7 shows the simulation waveform of testbench of Example 7.6, with both simulation waveforms of Figs. 7.3, 7.5, and 7.7 showing the same functionality.

The schematic of Example 7.5 is similar to the schematic of Example 7.1 shown in Fig. 7.4. The RTL coding for the state machine shown in Examples 7.1, 7.3, and 7.5 has the same functionality and synthesizes to similar logic although they differ in coding style.

When to Use One-Hot, Gray, or Binary Encoding

There are three different types of state machine encoding. Table 7.1 shows how each encoding is differently represented. Gray encoding and binary encoding require n flip-flops for 2^n states. A state machine with 16 states requires 4 state flip-flops if using gray encoding or binary encoding. One-hot encoding uses one state flip-flop for each state bit. A state machine with 16 states requires 16 state flip-flops for one-hot encoding.

One-hot encoding is larger compared to gray encoding or binary encoding as it requires more state flip-flops. However one-hot encoding is faster compared to gray encoding or binary encoding, as only one state bit is used to describe state, thus a simpler logic as compared to binary or gray encoding that requires larger logic to decode the states.

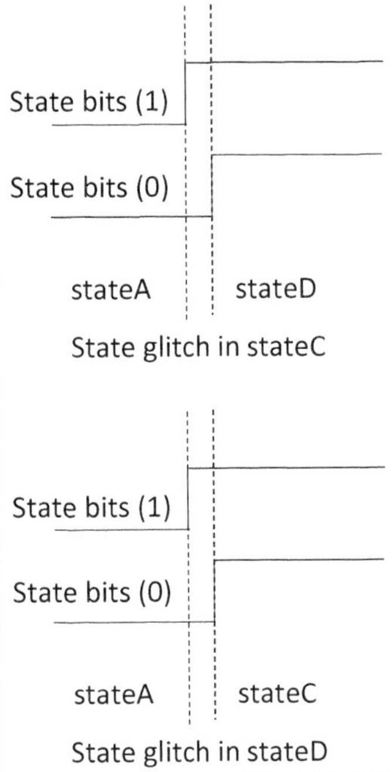

State	State bits (1:0) - binary
stateA	00
stateB	01
stateC	10
stateD	11

State	State bits (1:0) - gray
stateA	00
stateB	01
stateC	11
stateD	10

Fig. 7.8 Diagram showing state glitches for gray encoding and binary encoding

One-hot encoding is also state glitch-free. Binary and gray encoding require more logic to decode the different state bits to determine the state. Due to the delay inherent in these state bit decodings, there will be state glitches when state signal changes as shown in Fig. 7.8. Gray encoding has less state glitch compared to binary encoding as state glitches only occur in gray encoding when more than one state bit changes. Gray encoding uses less power to decode compared to binary encoding as adjacent states only change with one state bit.

A state machine design that requires fast timing or is state glitch-free will need to be designed using one-hot encoding. However, its area will be larger compared to one-hot or gray encoding. A state machine that requires less space as compared to one hot, less power for decoding state bits, and less state glitches when moving between adjacent states is suitable for gray encoding.

Blocking Statements

The RTL Verilog shown in Examples 7.1, 7.3, and 7.5 uses <= and = in some of the signal generation. The Verilog statements using <= are referred to as non-blocking statements.

```
        if (reset)
                present_state <= IDLE;
        else
                present_state <= next_state;
```

The Verilog statements using = are referred to as blocking statements.

```
        if (reset)
                ns = IDLE;
        else
        begin
                case (present_state)
                        IDLE:
                                begin
                                if (validfier1 & validfier2)
                                        ns = STATE1;
                                else if (validfier1 & validfier3)
```

Blocking statements and non-blocking statements are used in procedural statements. Typically blocking statement and non-blocking statements refer to signal generation using = and < = in always block or initial block.

Procedural statements are sequential statements whereby a statement is executed before moving to the next statement.

Example 7.7 Example showing Verilog code of blocking statement

```
always @ (A or B or C or Q1 or Q2)
begin
        Q1 = A & B;
        Q2 = Q1 & C;
        Q3 = Q1 & Q2 & B & C;
end
```

The RTL code shown in Example 7.7 uses blocking statement. The three statements are sequentially executed with one after the other. Q1 = A & B; is executed before moving to executing Q2 = Q1 & C; and finally executing Q3 = Q1 & Q2 & B & C;. A change in sequence of the statements as shown in Example 7.8 changes the order of the sequential statement execution and results in a different behavior of the RTL.

Example 7.8 Example showing Verilog code of blocking statement with change in sequence of statement

```
always @ (A or B or C or Q1 or Q2)
begin
```

```
        Q3 = Q1 & Q2 & B & C;
        Q1 = A & B;
        Q2 = Q1 & C;
end
```

Referring to Example 7.8, Q3 = Q1 & Q2 & B & C; is executed with previous value of Q1 and Q2 being used to generate Q3. Next is Q1 = A & B; being executed with Q1 being updated in this statement. The execution of Q2 = Q1 & C; will use the updated Q1 value. Example 7.9 shows the RTL code using the sequence shown in Example 7.7 and a testbench to simulate the RTL code.

Example 7.9 Example showing Verilog code of blocking statement

```
module blocking (A, B, C, Q1, Q2, Q3);
input A, B, C;
output Q1, Q2, Q3;
reg Q1, Q2, Q3;
always @ (A or B or C or Q1 or Q2)
begin
        Q1 = A & B;
        Q2 = Q1 & C;
        Q3 = Q1 & Q2 & B & C;
end
endmodule
module blocking_tb ();
reg A, B, C;
wire Q1, Q2, Q3;
integer i;
initial
begin
        {A, B, C} = 0;
        for (i=0; i<8; i=i+1)
        begin
                {A, B, C} = i;
                #10;
        end
end
blocking blocking_inst (.A(A), .B(B), .C(C), .Q1(Q1), .Q2(Q2),
.Q3(Q3));
endmodule
```

Figure 7.9 shows the simulation results of the testbench of Example 7.9 to simulate the sequence of blocking statements as shown in Example 7.7. Figure 7.10 shows the simulation results of the same testbench but with the sequence of block-

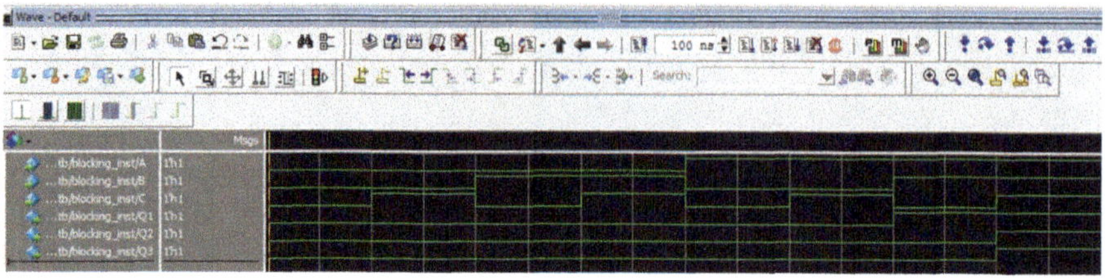

Fig. 7.9 Diagram showing simulation results of blocking statement using sequence shown in Example 7.7

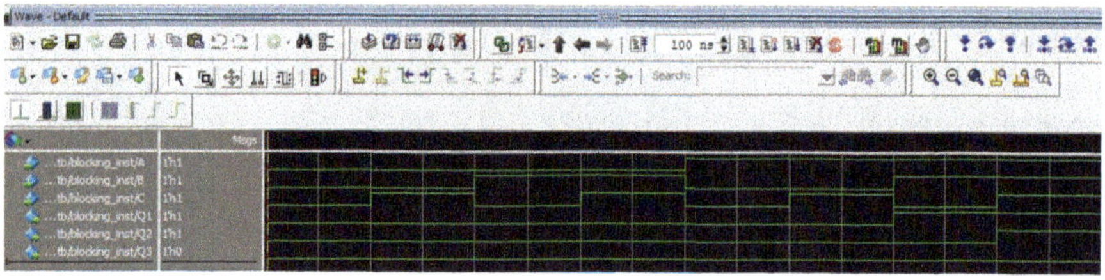

Fig. 7.10 Diagram showing simulation results of blocking statement using sequence shown in Example 7.8

ing statements as shown in Example 7.8. Comparing both the simulation results, Q3 has a different behavior. From Fig. 7.9:

(a) When A, B, and C are at logic 1, Q1 = A & B goes high.
(b) As Q1 is high from (a) and C is high, Q2 = Q1 & C goes high.
(c) As Q1 from (a) and Q2 from (b) are high, Q3 = Q1 & Q2 & B & C goes high.

From Fig. 7.10:

(a) When A, B, and C are at logic 1, Q3 = Q1 & Q2 & B & C uses Q1 and Q2 from previous value. The previous value of Q1 is high and Q2 is low; thus Q3 remains low.
(b) Q1 = A & B goes high as both A and B are high.
(c) As Q1 is high from (b) and C is high, Q2 = Q1 & C goes high.

The simulation results of Figs. 7.9 and 7.10 show that the change of sequence of the blocking statement in the always block will change the behavior of the RTL code.

Non-blocking Statements

Non-blocking statements used in procedural statements use <=. When a sequential statement uses a non-blocking statement, the assignment of the statement does not occur immediately. All the non-blocking statement in an always block is only executed at the end of the simulation cycle when all the statements within the always block are evaluated.

Using the same Verilog sequence of Examples 7.7 and 7.8, the Verilog sequence is modified using non-blocking statement as shown in Examples 7.10 and 7.11.

Example 7.10 Example showing Verilog code of non-blocking statement

```
always @ (A or B or C or Q1 or Q2)
begin
        Q1 <= A & B;
        Q2 <= Q1 & C;
        Q3 <= Q1 & Q2 & B & C;
end
```

The RTL code shown in Example 7.10 uses non-blocking statement. Although the three non-blocking statements are in an always block, they are only executed at the end of simulation cycle when all the three non-blocking statements are evaluated. The first non-blocking statement Q1 < = A & B; is executed together with Q2 < = Q1 & C; and Q3 < = Q1 & Q2 & B & C; when all three non-blocking statements are evaluated. Example 7.11 shows a change in sequence of the statements on the order of the sequential statement execution.

Example 7.11 Example showing Verilog code of non-blocking statement with change in sequence of statement

```
always @ (A or B or C or Q1 or Q2)
begin
        Q3 <= Q1 & Q2 & B & C;
        Q1 <= A & B;
        Q2 <= Q1 & C;
end
```

Referring to Example 7.11, all three non-blocking statements are executed together when all three statements have been evaluated. Example 7.12 shows the RTL code using the non-blocking sequence shown in Example 7.11 and a testbench to simulate the RTL code.

Example 7.12 Example showing Verilog code of non-blocking statement

```
module nonblocking (A, B, C, Q1, Q2, Q3);
input A, B, C;
output Q1, Q2, Q3;
reg Q1, Q2, Q3;
always @ (A or B or C or Q1 or Q2)
begin
```

```
    Q3 <= Q1 & Q2 & B & C;
    Q1 <= A & B;
    Q2 <= Q1 & C;
end
endmodule
module nonblocking_tb ();
reg A, B, C;
wire Q1, Q2, Q3;
integer i;
initial
begin
    {A, B, C} = 0;
    for (i=0; i<8; i=i+1)
    begin
        {A, B, C} = i;
        #10;
    end
end
nonblocking nonblocking_inst (.A(A), .B(B), .C(C), .Q1(Q1), .
Q2(Q2), .Q3(Q3));
endmodule
```

Figure 7.11 shows the simulation results of the testbench of Example 7.12 to simulate the sequence of non-blocking statements as shown in Example 7.10. Figure 7.12 shows the simulation results of the same testbench but with the sequence of non-blocking statements as shown in Example 7.11. Both have the same simulation results. From Fig. 7.12:

(a) When A, B, and C are at logic 1, Q1 < = A & B goes high.
(b) As Q1 is executed together with Q2 and C is at logic 1, Q2 < = Q1 & C goes high.
(c) As Q3 is executed together with Q1 and Q2, and B, C is at logic 1, Q3 < = Q1 & Q2 & B & C goes high.

From Fig. 7.11:

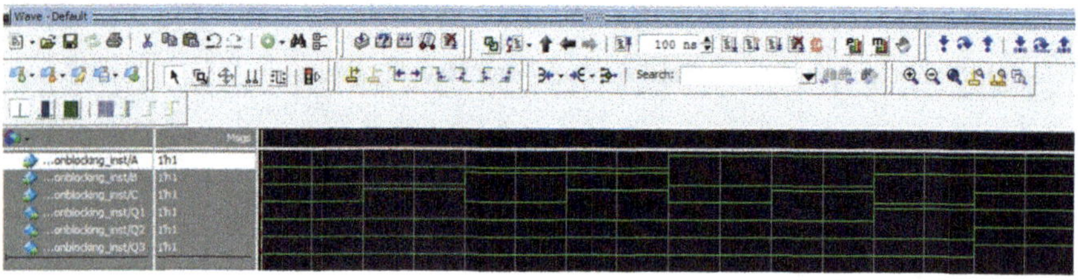

Fig. 7.11 Diagram showing simulation results of non-blocking statement using sequence shown in Example 7.10

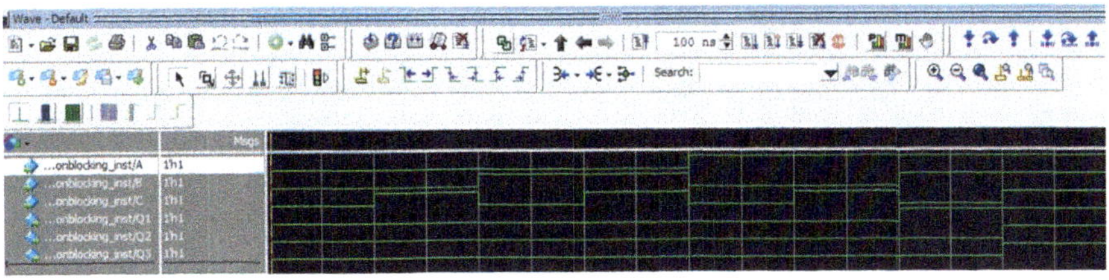

Fig. 7.12 Diagram showing simulation results of non-blocking statement using sequence shown in Example 7.11

(a) When A, B, and C are at logic 1, Q3 < = Q1 & Q2 & B & C goes high as Q1 and Q2 are executed together with Q3.
(b) When C is at logic high and Q1 at logic 1, Q2 < = Q1 & C goes high as all three non-blocking statements are executed together.
(c) Q1 < = A & B goes high when both A and B are high.

Referring to the simulation results of Figs. 7.11 and 7.12, the simulation results are the same irrespective of the sequence of non-blocking statements. This is different from blocking statements as blocking statements are executed sequentially. However, non-blocking statements are executed together when all the statements in the always block are evaluated. Thus a change in sequence of non-blocking statement does not affect its simulation functionality.

Rule of Thumb when Using Non-blocking Statement and Blocking Statement

As non-blocking statement in an always block is executed together at the end of the simulation timestep when all the statements are evaluated, a change in sequence of order of the non-blocking statement does not affect the behavior of the RTL.

Blocking statement in an always block is executed sequentially, thus making them truly sequential. Therefore a change in sequence of blocking statements in an always block will affect the behavior of the RTL.

When writing RTL Verilog for synthesis, a rule of thumb to observe to ensure simulation results match synthesis results is as follows:

(a) When writing RTL code for sequential logic or flip-flops, use non-blocking statement.
    (i) Figure 7.13 Shows the schematic diagram of a three-flip-flop register.
    (ii) The RTL Verilog code for the three-flip-flop design must use non-blocking statement as shown in Example 7.13.

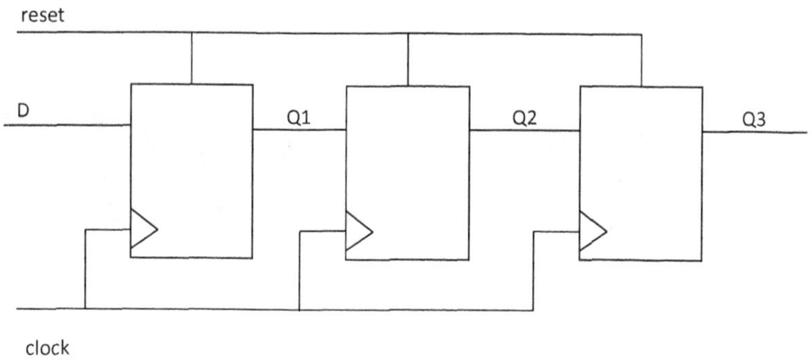

Fig. 7.13 Diagram showing schematic of a three-flip-flop register

Example 7.13 Example showing Verilog code of non-blocking statement for sequential logic

```
module flipflopnonblocking (D, clock, reset, Q3);
input D, clock, reset;
output Q3;
reg Q1, Q2, Q3;
always @ (posedge clock or posedge reset)
begin
        if (reset)
                begin
                        Q1 <= 0;
                        Q2 <= 0;
                        Q3 <= 0;
                end
        else
                begin
                        Q1 <= D;
                        Q2 <= Q1;
                        Q3 <= Q2;
                end
end
endmodule
```

(iii) By using non-blocking statement for sequential logic, the sequence of non-blocking statement in the always block will not affect the simulation results.

(b) When writing RTL code for combinational logic, use blocking statement.

(i) Figure 7.14 Shows the schematic diagram of a combinational logic with the truth table of its functionality.

(ii) The RTL Verilog code for the combinational logic must use blocking statement as shown in Example 7.14.

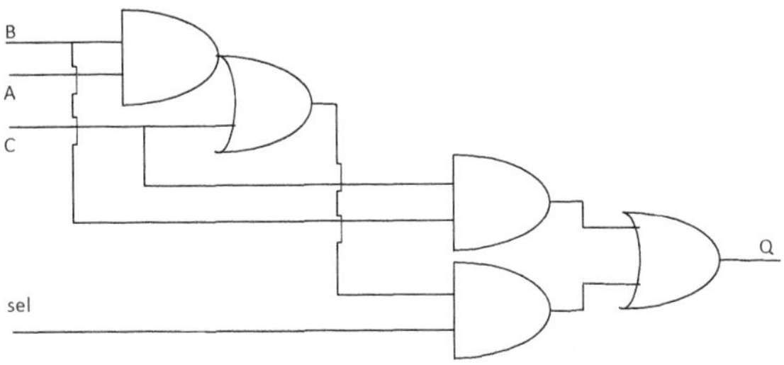

sel	A	B	C	Q
0	0	0	0	0
0	0	0	1	0
0	0	1	0	0
0	0	1	1	1
0	1	0	0	0
0	1	0	1	0
0	1	1	0	0
0	1	1	1	1
1	0	0	0	0
1	0	0	1	1
1	0	1	0	0
1	0	1	1	1
1	1	0	0	0
1	1	0	1	1
1	1	1	0	1
1	1	1	1	1

Fig. 7.14 Diagram showing schematic of a combinational logic with its truth table

Example 7.14 Example showing Verilog code of blocking statement for combinational logic

```
module combiblocking (A, B, C, sel, Q);
input A, B, C, sel;
output Q;
reg Q;
always @ (A or B or C or sel)
begin
        if (sel)
                Q = A & B | C;
        else
                Q = B & C;
end
endmodule
```

(iii) Sequential statement within an always block to represent the logic functionality of combinational logic uses blocking statement as blocking statement is truly sequential, thus correctly representing the simulation results of the combinational logic.

By following these two rules in RTL coding, mismatch between simulation and synthesis due to usage of blocking statement and non-blocking statement can be avoided. Basically in RTL, sequential logic uses non-blocking statement in always block, while combinational logic uses blocking statement in always block.

Chapter 8
RTL Coding Guideline

When writing RTL Verilog for synthesis, it is important to follow certain set of rules for the RTL coding. This ensures that the RTL code is synthesizable and does not create any unnecessary mismatch between RTL simulation and gate level simulation. It is important to note that RTL that simulates correctly does not necessary synthesizes correctly. The following are several rules to follow during RTL coding to ensure the RTL Verilog code synthesizes correctly.

Contention

Contention occurs in RTL Verilog code if there are two separate statements driving the same signal. For always block, if two separate always block drive the same signal, this will create contention in synthesis as each always block synthesizes to a certain logic, with two logic blocks driving the same signal (Fig. 8.1).

Simulators typically do not flag an error when there is contention in a Verilog code. However, synthesis tool commonly will flag an error when the RTL Verilog code has contention. Example 8.1 shows a Verilog code with contention on internal signal X1.

Example 8.1 Example showing Verilog code with contention on two always blocks driving the same signal

```
module contentian (A, B, C, sel, Q1, Q2);
input A, B, C, sel;
output Q1, Q2;
reg X1, X2;
reg Q1, Q2;

always @ (A or B)
```


W. F. Lee, *Learning from VLSI Design Experience*,
https://doi.org/10.1007/978-3-030-03238-8_8

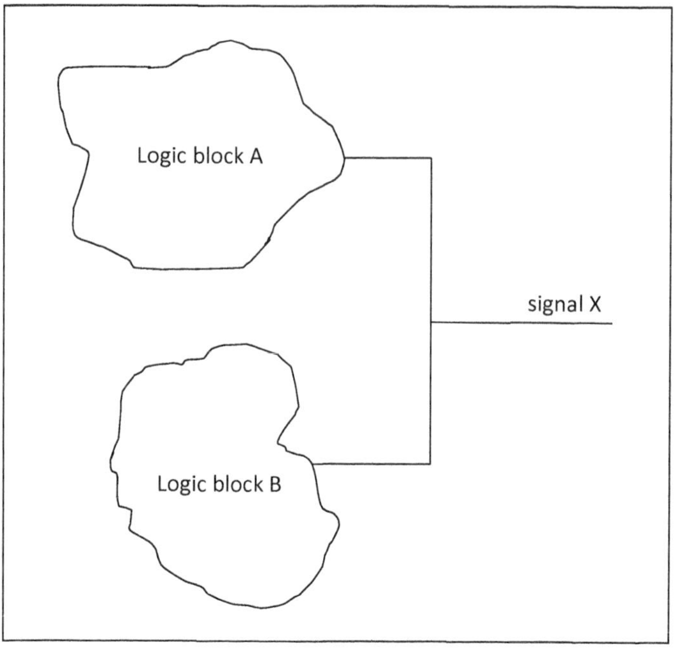

Fig. 8.1 Diagram showing two logic blocks driving the same signal

```
begin
        X1 = A & B;
        X2 = A | B;
end

always @ (B or C or X1 or X2)
begin
        if (sel)
                X1 = B & C;
        else
                X1 = X2;
        Q1 = X1 | C;
        Q2 = X2 | C;
end

endmodule
```

Referring to Example 8.1, both always blocks are driving the signal X1, thus causing a contention. Example 8.2 shows the same Verilog code but using assign statements, with two assign statement driving the same signal X1 causing contention.

Example 8.2 Example showing Verilog code with contention on two assign statements driving the same signal

```
module contentionassign (A, B, C, sel, Q1, Q2);
input A, B, C, sel;
output Q1, Q2;
wire X1, X2;
wire Q1, Q2;

assign X1 = A & B;
assign X2 = A | B;

assign X1 = sel ? (B&C) : X2;
assign Q1 = X1 | C;
assign Q2 = X2 | C;

endmodule
```

Sensitivity List

Sensitivity list refers to the list of signals specified in an always block.

```
always @ (A or B or C)
```

Whenever a signal in the sensitivity list changes, the statements inside the always block will be evaluated. For Verilog 1995 standard, the sensitivity list is separated by keyword *or*. For Verilog 2001 standard, the sensitivity list can be separated by, or by using *.

```
always @ (A, B, C)
always @ *
```

However some synthesis tools do not accept Verilog 2001 and requires the RTL Verilog code to follow Verilog 1995 using keyword *or* to separate signals in the sensitivity list.

Sensitivity list play an important role in simulation as change in the signals of a sensitivity list will evaluate the statements in the always block. Therefore if a RTL Verilog code has an incomplete sensitivity list, RTL simulation results may not match gate level simulation.

Example 8.3 shows a Verilog RTL code of a logic functionality. The sensitivity list of the always block is intentionally left incomplete without signal C.

Example 8.3 Example showing RTL Verilog Code with incomplete sensitivity list

```
module incompletesensitivity (A, B, C, sel, Q1);
input A, B, C;
input [1:0] sel;
output Q1;
reg Q1;

always @ (sel or A or B)
begin
        case (sel)
                2'b00: Q1 = A & B & C;
                2'b01: Q1 = A & B | C;
                2'b10: Q1 = A | B & C;
                2'b11: Q1 = A | B | C;
                default: Q1 = 0;
        endcase
end
endmodule
```

Example 8.4 shows a Verilog testbench to pump stimulus into the RTL Verilog code of Example 8.3 with incomplete sensitivity list.

Example 8.4 Example showing Verilog testbench to simulate the incomplete sensitivity list of RTL Verilog code

```
module incompletesensitivity_tb ();
reg A, B, C;
reg [1:0] sel;
wire Q1;
integer i, j;

initial
begin
        {A, B, C, sel} = 0;
        #10;
        for (i=0; i<4; i=i+1)
        begin
                for (j=0; j<8; j=j+1)
                begin
                        sel = i;
                        {A, B, C} = j;
                        #10;
                end
        end
```

```
end

incompletesensitivity inst (.A(A), .B(B), .C(C), .sel(sel), .
Q1(Q1));

endmodule
```

Figure 8.2 shows the simulation results of Verilog testbench, Example 8.4.

Referring to Fig. 8.2, as signal C is not listed in the sensitivity list, thus change in signal C does not trigger the always block and the Verilog statements in the always block are not evaluated. The changes in output Q1 occur when signal changes for signals A, B, and sel, thus triggering the sensitivity list and evaluating the Verilog statements in the always block.

Sensitivity list is not considered in synthesis; thus incomplete sensitivity list does not affect synthesis although most synthesis tools will give a warning message that there are signals missing from a sensitivity list. Example 8.5 shows the gate level netlist generated from the Mentor Graphics' Leonardo Spectrum synthesis tool after it synthesizes the RTL code of Example 8.3.

Example 8.5 Example showing Verilog gate level netlist of RTL Verilog with incomplete sensitivity list

```
module incompletesensitivity ( A, B, C, sel, Q1 ) ;
    input A ;
    input B ;
    input C ;
    input [1:0]sel ;
    output Q1 ;
    wire nx8, nx99;
    ao221 ix15 (.Y (Q1), .A0 (sel[1]), .A1 (A), .B0 (sel[0]), .B1
(C), .C0 (nx8)) ;
    inv01 ix9 (.Y (nx8), .A (nx99)) ;
    oai221 ix100 (.Y (nx99), .A0 (C), .A1 (sel[0]), .B0 (A), .B1
(sel[1]), .C0 (B)) ;
endmodule
```

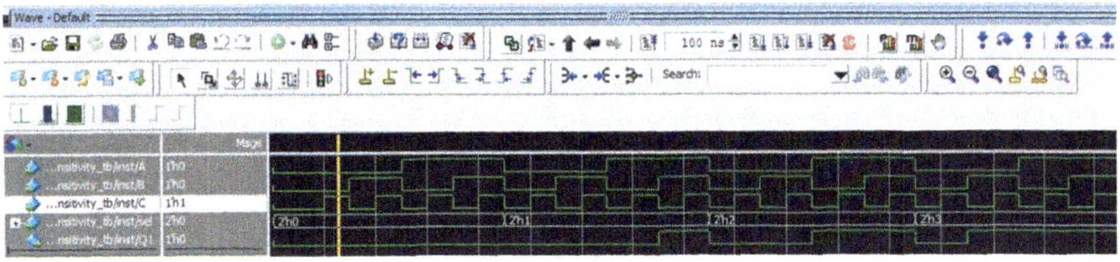

Fig. 8.2 Diagram showing simulation results of incomplete sensitivity list

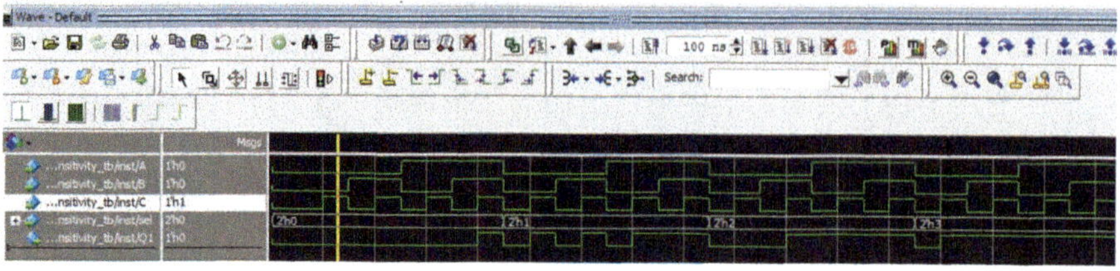

Fig. 8.3 Diagram showing gate level simulation results of synthesized netlist with incomplete sensitivity list

Using the same testbench shown in Example 8.4, Fig. 8.3 shows the simulation results of gate level simulation of the synthesized netlist shown in Example 8.5.

Referring to Fig. 8.3, the gate level simulation results is different from the RTL simulation results shown in Fig. 8.2. The gate level simulation shows the output signal Q1 changes when C changes as opposed to the RTL simulation results shown in Fig. 8.2.

When writing RTL Verilog for synthesis, it is important to ensure the sensitivity list is complete to avoid mismatch between gate level simulation and RTL simulation.

Level-Sensitive and Edge-Sensitive RTL

Edge-Sensitive RTL Verilog Code

When using always block in RTL Verilog code, the always block can be used to describe the functionality of sequential logic or combinational logic. Sequential logic refers to flip-flops which are edge-sensitive registers as shown in Fig. 8.4.

Figure 8.4 shows the basic data flip-flop which is an edge-sensitive sequential logic. It is represented in Verilog as shown in Example 8.6.

Example 8.6 Example showing RTL Verilog for an edge-sensitive data flip-flop

```
module dataflop (D, clock, Qpos, Qneg);
input D, clock;
output Qpos, Qneg;
reg Qpos, Qneg;

always @ (posedge clock)
begin
        Qpos <= D;
end
```

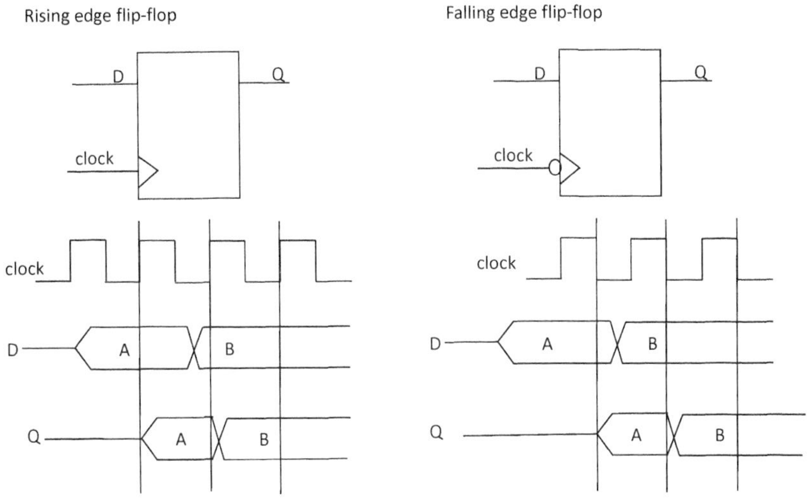

Fig. 8.4 Diagram showing edge-sensitive flip-flop

```
always @ (negedge clock)
begin
        Qneg <= D;
end

endmodule
```

The RTL Verilog code in Example 8.6 is a data flip-flop. It is common to use flip-flop with reset for design. There are asynchronous reset flip-flop and synchronous reset flip-flop. Synchronous reset flip-flop only recognizes the reset signal when the clock is at rising edge for a rising edge flip-flop. Asynchronous reset flip-flop recognizes the reset condition whenever reset signal is valid irrespective of clock. Figure 8.5 shows the functional waveform of a synchronous reset flip-flop and an asynchronous flip-flop.

Referring to Fig. 8.5, for asynchronous reset flip-flop, the output Q is low whenever reset goes high irrespective of clock. For synchronous reset flip-flop, output Q is low when reset goes high and clock is at rising edge. Example 8.7 shows the RTL Verilog code for an asynchronous reset rising edge flip-flop and a synchronous reset rising edge flip-flop.

Example 8.7 Example showing RTL Verilog for asynchronous reset and synchronous reset rising edge flip-flop

```
module resetflop (D, clock, reset, syncresetQ, asyncresetQ);
input D, clock, reset;
output syncresetQ, asyncresetQ;
```

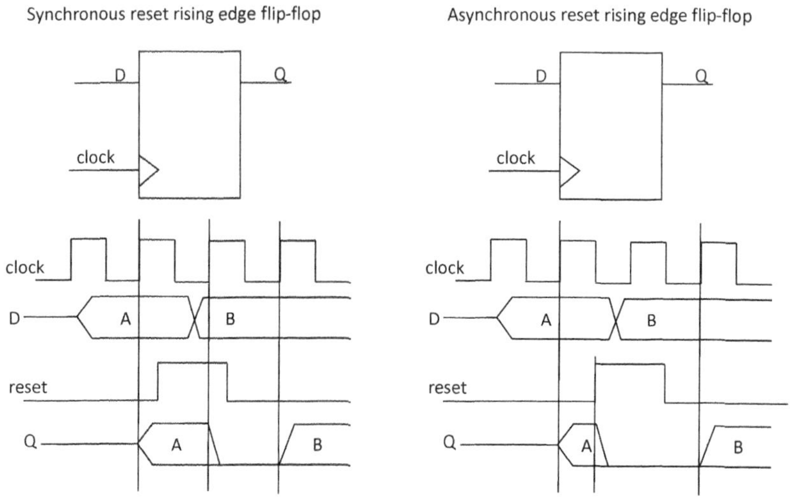

Fig. 8.5 Diagram showing asynchronous reset and synchronous reset edge-sensitive flip-flop

```
reg asyncresetQ, syncresetQ;

always @ (posedge clock or posedge reset)
begin
        if (reset)
                asyncresetQ <= 0;
        else
                asyncresetQ <= D;
end

always @ (posedge clock)
begin
        if (reset)
                syncresetQ <= 0;
        else
                syncresetQ <= D;
end
endmodule
```

Similar to asynchronous reset flip-flop and synchronous reset flip-flop, a sequential logic can also be a synchronous set flip-flop or asynchronous set flip-flop. A functionally more complex version of sequential element is an asynchronous set reset flip-flop. Example 8.8 shows the RTL code for an asynchronous set reset flip-flop. The output of flip-flop goes low when reset goes high or the output of the flip-flop goes high when set goes high.

Example 8.8 Example showing RTL Verilog for asynchronous reset and asynchronous set rising edge flip-flop

```
module setresetflop (D, clock, reset, set, Q);
input D, clock, reset, set;
output Q;
reg Q;

always @ (posedge clock or posedge reset or posedge set)
begin
        if (reset)
                Q <= 0;
        else if (set)
                Q <= 1;
        else
                Q <= D;
end
endmodule
```

Synthesis of the RTL Verilog code shown in Example 8.8 will return an error if the standard cell library used for the synthesis does not have a flip-flop with asynchronous set and asynchronous reset. However, most standard cell library have many different types of flip-flop allowing for the synthesis to translate the RTL Verilog code to standard cell gates.

Level-Sensitive RTL Verilog Code

Combinational logic can be described in RTL Verilog using assign statements or procedural statements in an always block. When using always block, the sensitivity list must use level-sensitive code to describe combinational logic.

Example 8.9 Example showing RTL Verilog for level-sensitive combinational logic

```
module combi (A, B, C, D, Q1, Q2);
input A, B, C, D;
output Q1, Q2;
wire Q1;
reg Q2;

assign Q1 = A ? (B&C) : (D|B);


always @ (A or B or C or D)
```

```
begin
        if (A | C)
                Q2 = B;
        else
                Q2 = D & B;
end
endmodule
```

When writing RTL Verilog code for combinational logic using level sensitive in always block, the RTL code must have complete sensitivity list to ensure simulation matches between RTL and gate level.

Mixing Level-Sensitive and Edge-Sensitive Verilog Code

Mixing level-sensitive and edge-sensitive Verilog code in an always block can be simulated, but it cannot be synthesized. It is a non-synthesizable Verilog code.

Example 8.10 Example showing Verilog code mixing edge sensitive and level sensitive in always block

```
module mixing (A, B, C, clock, Q1);
input A, B, C, clock;
output Q1;
reg Q1;

always @ (posedge clock or A or B or C)
begin
        if (A & B)
                Q1 <= C;
        else if (~A)
                Q1 <= 0;
end
endmodule
```

The Verilog code shown in Example 8.10 is not an RTL code. It can simulate in Verilog simulator, but it will not synthesize in synthesis tool. The synthesis tool will give an error message of mixing edge-sensitive and level-sensitive trigger in the always block.

Input, Output, and Bidirectional Ports in RTL

All the Verilog examples shown uses only input ports and output ports. Verilog allows for bidirectional ports using the keyword "inout."

```
inout [7:0] bididata; // 8 bits bidirectional port
inout bidictrl; // single bidirectional port
```

However, when writing RTL Verilog, it is advisable to avoid using bidirectional ports or inout. Having inout in RTL complicates the control logic and timing constraints for synthesis and auto place & route. These complications arise as control logic is needed to ensure the bidirectional port behave as an output port when it is output and behave as an input port when it is input. The timing constraints for synthesis and auto place & route also need to specify required time when the port acts as an output port and arrival time when the port acts as an input port.

To avoid the complications arising from using bidirectional ports, it is common design rule to write RTL Verilog using only input port and output port. If there is any bidirectional port, it is only connected as a bidirectional port at the fullchip stage.

Referring to Fig. 8.6, the digital core RTL uses only input port and output port; bidirectional port is not allowed in RTL. At the final stage of fullchip layout, the input ports of digital core are connected to the input buffers. The output ports of digital core are connected to the output buffer. For bidirectional pins, the input and output are connected accordingly to the digital core as shown in Fig. 8.6 but with an additional enable signal for controlling the output enable for the bidirectional IO.

Blocking and Non-blocking Statement

Blocking and non-blocking statements refer to the use of <= or = in signal assignments for procedural statements such as those in an always block.

Blocking statements are truly sequential as each statement is evaluated before moving to the next statement. Therefore, changing of sequence of blocking statements within an always block may change the simulation behavior as shown in Chap. 7.4.

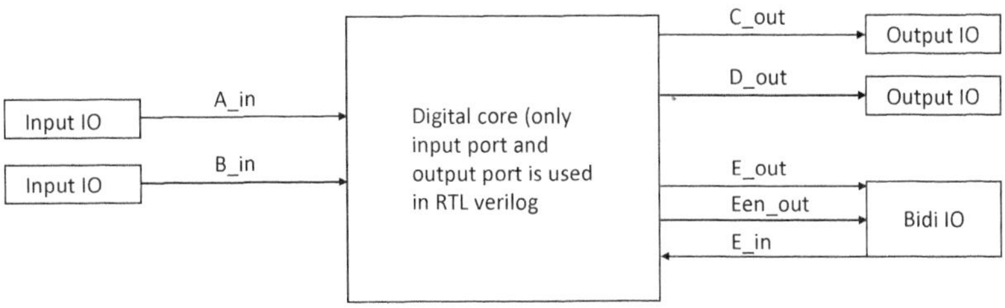

Fig. 8.6 Diagram showing fullchip connectivity for bidirectional IO

Non-blocking statements are only evaluated at the end of the simulation timestep, thus allowing changing of sequence within always block and not affecting the simulation behavior as shown in Chap. 7.5.

A safe rule of thumb, to write RTL Verilog code is to use non-blocking statements when using always block for sequential logic and use blocking statements when using always block for combinational logic. By adhering to this rule, the RTL Verilog code will always have a consistent simulation behavior irrespective of sequence of non-blocking statements.

Inferred Latch

An unwanted latch will be inferred in a design if an if-else statement or case statement is incomplete as shown in Chap. 4. For if-else statements, if there are four different possible conditions and only three are specified with the fourth condition not specified, the simulator will assume the value remains unchanged for the unspecified condition. This results in a latch inference. Similarly for case statements that do not have all conditions specified, the simulator assumes the value remain unchanged for the unspecified condition, thereby inferring a latch.

Inferred latches are unwanted and increases gate count, increases timing, and complicates DFT. To ensure if-else statements do not infer unwanted latches, it is good coding practice to always use the "else" statement to ensure all conditions are covered. Similarly for the case statement, it is good coding practice to always use the "default" statement to ensure all conditions are covered for the case statement to avoid unwanted latch inference.

Signed and Unsigned

When writing RTL Verilog code for signed calculations, mixing of signed and unsigned in a statement will result in the results of the calculation being treated as unsigned. It is important to note that mixing of signed and unsigned will result in the functionality being unsigned even though the functionality is declared as signed.

Logic between Blocks

An ASIC or SoC have many functionality with each similar or related functionality grouped into a logic block or module. The top-level design consists of multiple blocks or multiple design modules. These blocks are glued together at top level to form the fullchip. Each logic block is commonly averaged at 200,000 gates or

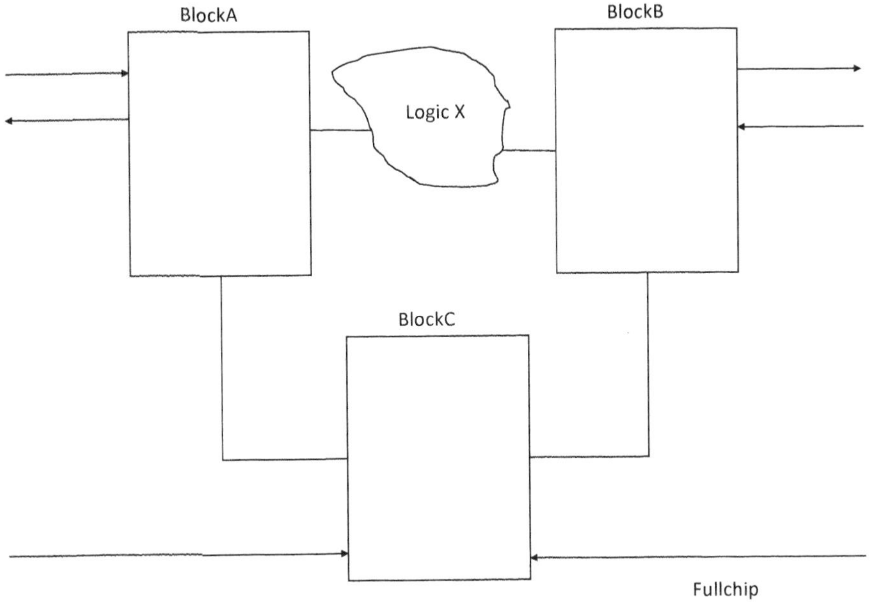

Fig. 8.7 Diagram showing top level glue logic between two blocks

thereabouts. A logic block with too few logic gates is inefficiently optimized during synthesis due to lack of logic sharing opportunities. A logic block that is too large will have the synthesis tool handling too much optimization resulting in inefficient synthesis.

At the top fullchip level where different logic blocks are glued together, logic between blocks should be avoided. The logic between these blocks cannot be efficiently optimized with the logic within the blocks as shown in Fig. 8.7. The glue logic X cannot be efficiently optimized with the logic within Block A or Block B as synthesis does not alter the hierarchy ports unless hierarchy is intentionally broken. Designers normally do not intentionally break a hierarchy as broken hierarchy makes debugging difficult as signals may be optimized away making it hard to debug.

A recommended solution would be to move logic X into either Block A or Block B to allow logic X to be optimized with the logic within those blocks.

Register Output of Blocks

When designing a logic block, the designer will assume an arrival time on the input signals and required time on the output signals. The output signals of a logic block will connect to another logic block as an input signal, in which it will require arrival time of the signal. An output signal of a block that connects to another block will be the input signal to that block. Thus, required time of a signal that connects to another block as input will be the arrival time for that block as shown in Fig. 8.8 with the added delay due to the interconnect.

Due to the arrival time for input signal of a block and required time for output signal of a block, designers will estimate their required time and arrival time based on information provided by designers of the block on which the signal is connected to. The estimation can be improved through time budgeting at the fullchip level. However, when in the early stages of design before a fullchip is achieved, designers must estimate these arrival time and required time for the block's timing constraints. A timing constraint that is too tight will result in the block being inefficient with many parallel paths to try to meet the tight timing, resulting in over-constrained. A timing constraint that is too loose will result in the block under optimized and having actual timing failures, resulting in under-constrained. Therefore, a good estimate on the arrival time and required time is crucial to obtain optimal design results.

A solution to resolve the inaccurate estimation of arrival time and required time is to register the output signals of the block as shown in Fig. 8.9.

Referring to Fig. 8.9, when the output of Block A is registered, the endpoint of the timing path remains within Block A. The startpoint and endpoint of the timing path for Block B remains within Block B. Similarly, the startpoint and endpoint of

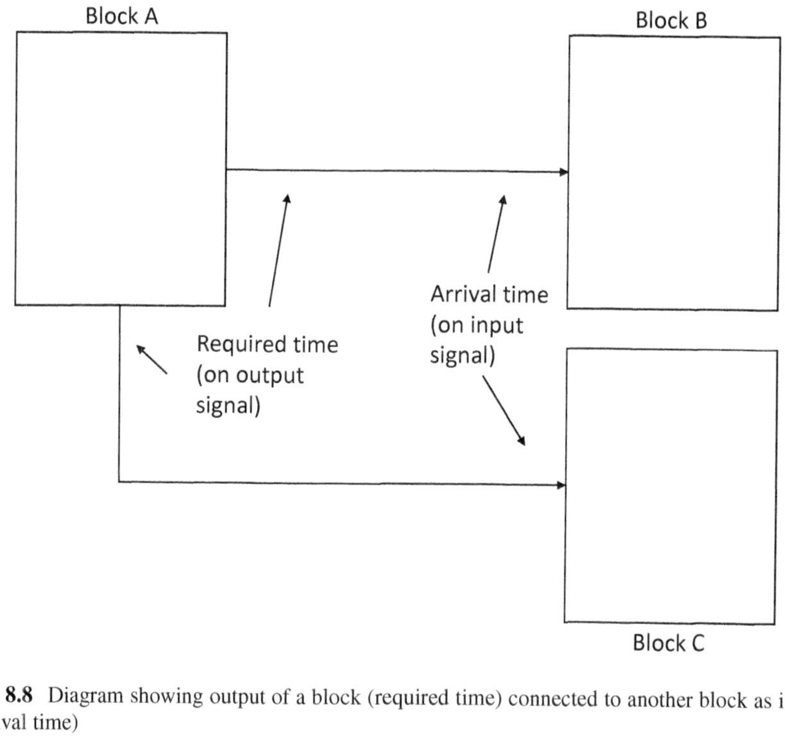

Fig. 8.8 Diagram showing output of a block (required time) connected to another block as input (arrival time)

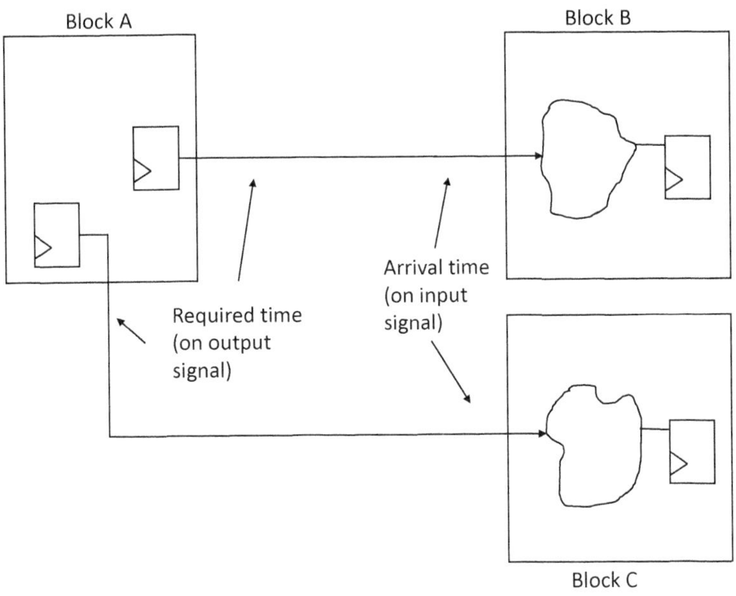

Fig. 8.9 Diagram showing output of a block being registered

the timing path for Block C remains within Block C. This solution removes the need to estimate the arrival time and required time as the path begins at the input of the block and ends at the register of each output within the block. Timing optimization is limited to within the block with better control on the timing paths.

Naming Convention

An ASIC and SOC commonly have millions of gates, a failure of a signal within a design is not easy to identify. However, if the design methodology includes naming convention as part of the methodology, the name of the failing signal can easily and quickly identify which module or logic block it is from and which module or logic block it is headed to.

Referring to Fig. 8.10, the naming convention shows:

(a) A2B_signal1—The name signal1 is from module A to module B.
(b) B2A_signal2B—The name signal2 is from module B to module A and active low.
(c) A2C_signal3—The name signal3 is from module A to module C.
(d) I2A_signal7—The name signal7 is from input to module A.
(e) A2O_signal8—The name signal8 is from module A to output.

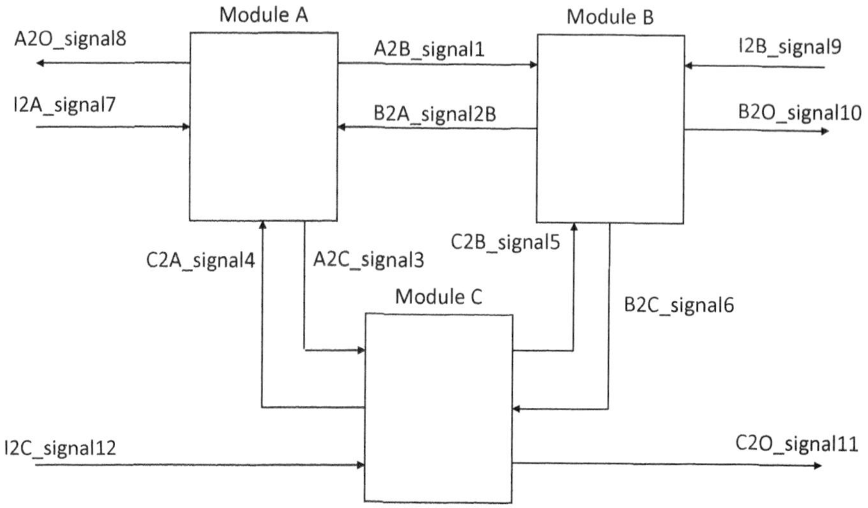

Fig. 8.10 Diagram showing naming convention for a design

The naming convention uses the module name of a signal source with the numerical number 2 and the module name of signal destination followed by the signal name. A “B” at the end of the signal name represents active low signal. If the signal is an input, it starts with “I” followed by “2” and the module name of the signal destination. If the signal is an output, it starts with the module name of the signal source followed by “2” and “O.”

By having such a naming convention, a failing signal can easily be identified its module source and module destination, whether it is a fullchip input or fullchip output, and if it is active low or active high.

Chapter 9
Code Coverage

Designers describe the functionality of a design using RTL code. The functionality of the RTL code is verified using a set of testbenches which pump in stimulus into the RTL code. The designer observes the behavior of the RTL code to identify any functionality that may not behave as expected. When such unexpected behavior occurs in the simulation, the designer fixes the RTL code and resimulates. This repeats in a loop until the designer is satisfied that the functionality of the RTL code is as expected.

A small RTL module may have several or tens of testbenches. But larger RTL modules may have 50 or more testbenches to simulate the different functionality and features of the RTL code. If a design does not have adequate testbenches to simulate its functionality, certain bugs may be present in the design that is not captured during verification. It is therefore important to ensure there are adequate testbenches to verify the different functionalities of the design.

A common method used in RTL verification is to monitor the simulation activity of the testbench on the RTL code and provides an indication on the percentage of the RTL code that has been exercised by the testbench. For a set of testbenches, each testbench simulation is monitored and the monitoring activity combined to provide a total percentage of RTL code that is exercised. The designer can decide if more testbenches are needed to exercise more of the RTL to gain a higher percentage. This process is commonly referred to as code coverage.

Flow for Code Coverage

Code coverage uses the RTL and testbenches to monitor the simulation and identifies which parts of the RTL have not been exercised. The flow in code coverage is similar to that for RTL simulation.

Figure 9.1 shows the code coverage flow in which the RTL code and testbenches are simulated. During simulation, the simulator monitors the simulation activity and

W. F. Lee, *Learning from VLSI Design Experience*,
https://doi.org/10.1007/978-3-030-03238-8_9

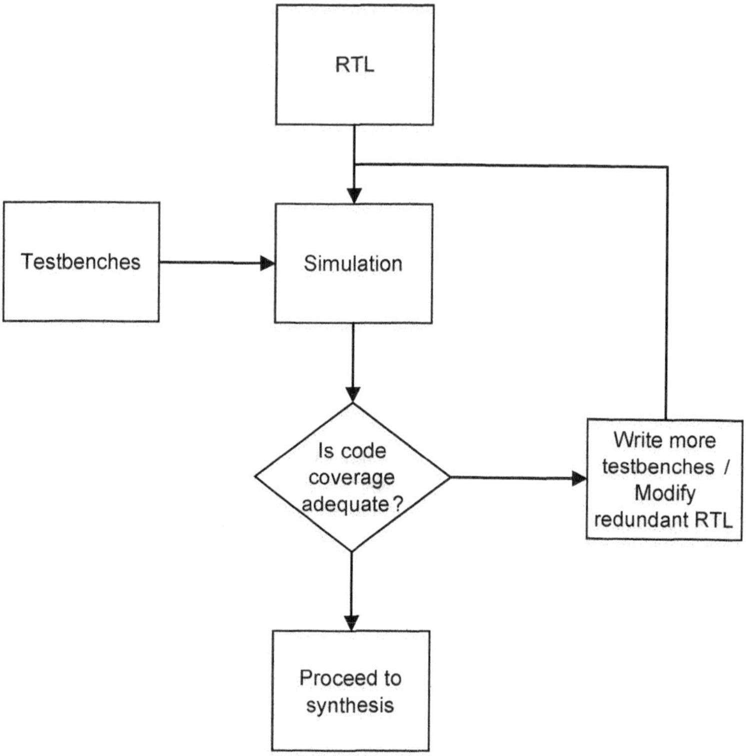

Fig. 9.1 Diagram showing flow for code coverage

identifies which parts of the RTL code are not exercised. If the percentage of the RTL code being exercised by the testbenches are too low, the designers must write more testbenches to increase the code coverage. Or the RTL code may have redundant code that may be removed.

Types of Code Coverage

Code coverage can be divided into several categories, supported by commonly used simulators such as Mentor Graphics' ModelSim:

1. Statement Coverage
   - This is the most basic code coverage in which a statement is being counted how many times it is executed.
2. Branch Coverage
   - Branch coverage is used in statements with if statement or case statement.

- Each statement in the branch must have taken its path.
- A branch which has not been executed will be highlighted.

3. Condition and Expression Coverage
   - Condition coverage is an extension of the branch coverage.
   - Expression coverage tracks the activity of the expression of a statement. It counts the execution of the right-hand side of assignment statements.
4. Toggle Coverage
   - Toggle coverage counts the changes of states on a node.
   - Toggle coverage can slow down a simulation significantly if data is collected for many nodes in a design.
5. Finite-State Machine Coverage
   - State machines are commonly used in design due to its predictability and synchronous nature.
   - Large state machines are complex due to its many transitions between states.
   - It is important to analyze the coverage of finite-state machine during RTL verification to ensure all state transitions are exercised.

Code coverage is a good indication if more testbenches are needed to verify a design, but good code coverage does not represent correct design functionality. Code coverage is merely a means to measure and indicate which parts of the RTL code are not exercised by the testbenches; it does not represent if the RTL code is functionally correct.

Simulation with Code Coverage

To have a closer look at code coverage, a Serial Peripheral Interface (SPI) is used as a design example with a testbench to simulate the functionality of the specified example. Code coverage is used on the testbench simulation to identify which parts of the RTL code are not exercised and how additional test conditions can be set up to increase the code coverage.

Serial Peripheral Interface (SPI) is a serial bus that was introduced by Motorola. It is widely used as it is simple and cost effective to implement. It is used in master/slave mode whereby a master initiates a transfer to/from a slave. There may be more than one slave on a SPI bus. Figure 9.2 shows a master slave configuration of a SPI interface using four wires: SCK, CSB, MOSI, and MISO.

Referring to Fig. 9.2:

a. SCK
   1. This is the SPI clock signal.
   2. The SPI slave uses the SCK signal as the SPI clocking reference.

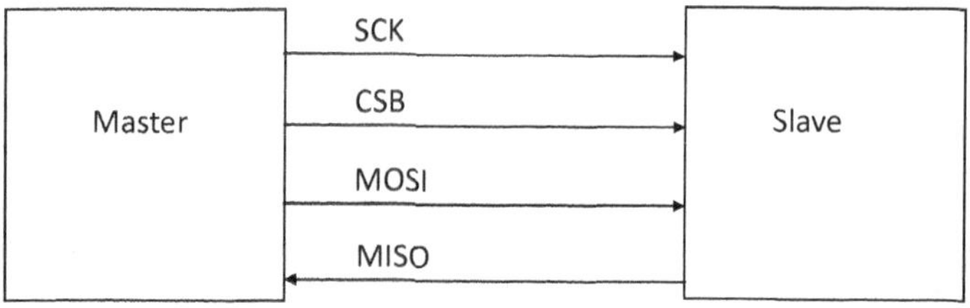

Fig. 9.2 Diagram showing four-wire SPI interface

b. CSB

1. This represents chip select active low.
2. When this signal is asserted, the SPI slave is enabled.

c. MOSI

1. Master out slave in. This signal represents data from master to slave.
2. It is a single-bit serial data.

d. MISO

1. Master in slave out. This signal represents data from slave to master.
2. It is a single-bit serial data.

The CSB signal can be used by the SPI master to control multiple different SPI slaves as shown in Fig. 9.3.

Referring to Fig. 9.3, the SPI master accesses each SPI slave through the CSxB signal. The MISO from the slaves are connected together; thus each SPI slave that is not accessed must tristate its MISO output.

Figure 9.4 shows the block diagram for a SPI slave which internally has four 8-bit register that a SPI master can access to write to and read from. The SPI signals are asynchronous to the clock signal of the SPI slave; thus clock domain crossing must be addressed at the MISO, CSB, and MOSI input signals.

The SPI master write to and read from the SPI slave is governed by the SPI command sent by the SPI master to the SPI slave. An advantage of SPI bus is its flexibility in specification of the SPI transaction. For the case of the block diagram shown in Fig. 9.4, the first 8 bits sent from SPI master to SPI slave is the SPI command to determine a read/write and to/from which of the four registers to access. Figure 9.5 shows the SPI bus protocol for the SPI command followed by the SPI data.

Referring to Fig. 9.5:

a. If D0[7]=0, this is a read from spi_slave.
b. If D0[1:0]=00, this is a read from register decelerate. D1[7:0] is data sent out by spi_slave.
c. If D0[1:0]=01, this is a read from register velocity. D1[7:0] is data sent out by spi_slave.

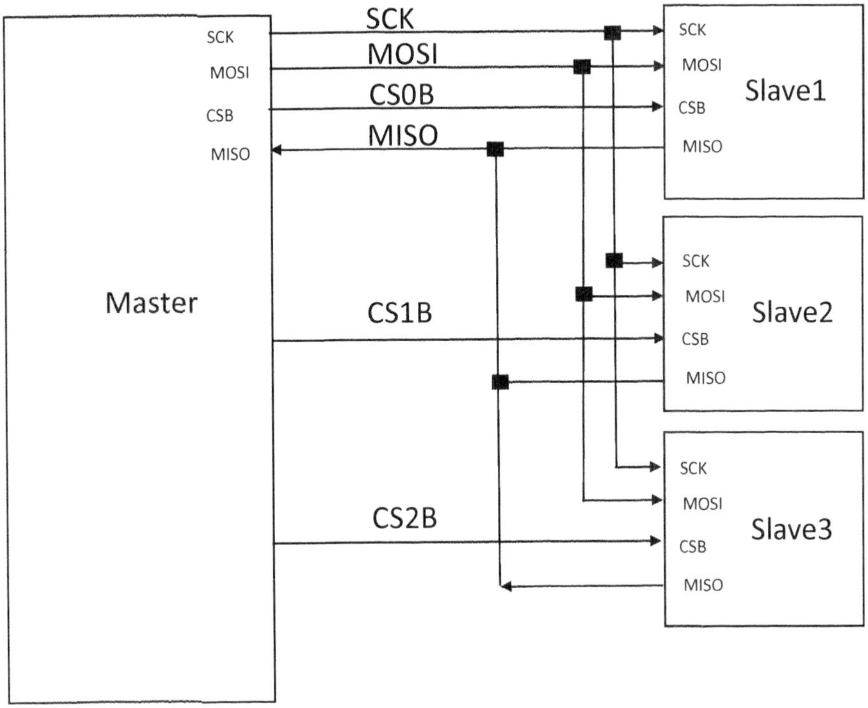

Fig. 9.3 Diagram showing SPI master control multiple SPI slaves

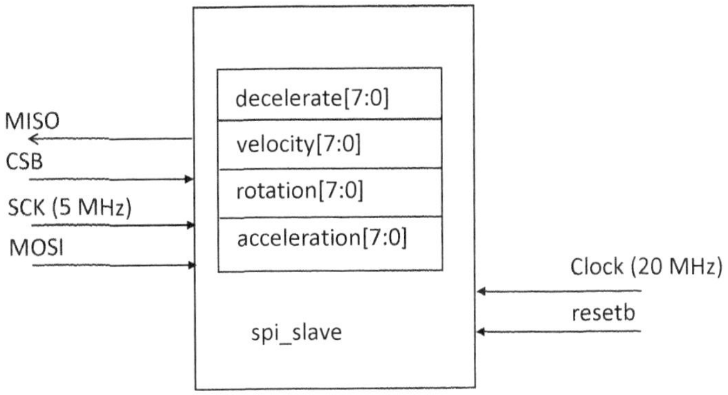

Fig. 9.4 Diagram showing SPI slave with four internal registers that SPI master can access

d. If D0[1:0]=10, this is a read from register rotation. D1[7:0] is data sent out by spi_slave.
e. If D0[1:0]=11, this is a read from register acceleration. D1[7:0] is data sent out by spi_slave.
f. D0[6:2] are don't care, not used for this particular example application.

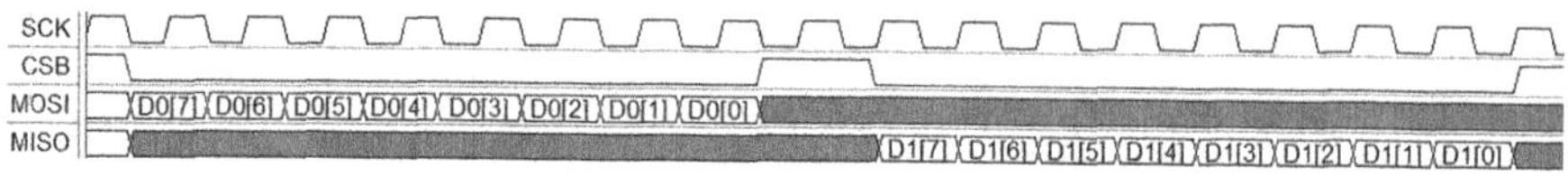

Fig. 9.5 Diagram showing SPI command protocol for SPI read

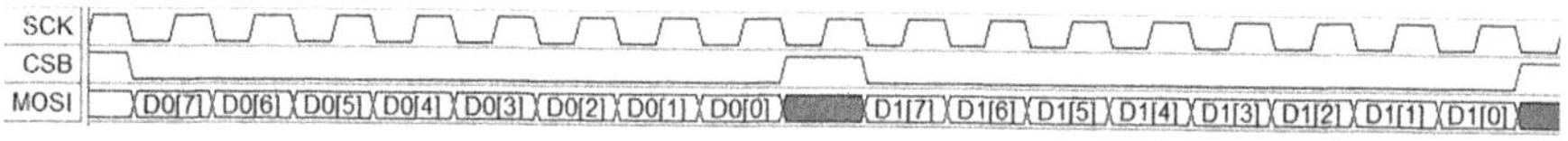

Fig. 9.6 Diagram showing SPI command protocol for SPI write

Referring to Fig. 9.6:

a. If D0[7]=1, this is a write to spi_slave.
b. If D0[1:0]=00, this is a write to register decelerate. D1[7:0] is data written to spi_slave.
c. If D0[1:0]=01, this is a write to register velocity. D1[7:0] is data written to spi_slave.
d. If D0[1:0]=10, this is a write to register rotation. D1[7:0] is data written to spi_slave.
e. If D0[1:0]=11, this is a write to register acceleration. D1[7:0] is data written to spi_slave.
f. Similar to SPI read, D0[6:2] are don't care for SPI write, not used for this particular example application.

Referring to the block diagram of Fig. 9.4 and the SPI protocol shown in Figs. 9.5 and 9.6, a block architecture can be derived as shown in Fig. 9.7, keeping in mind a clock domain crossing synchronizer is needed at the incoming SPI signals.

The SPI input signals SCK, CSB, and MOSI go through a CDC synchronizer. An edge detector logic detects the SCK edge as well as the CSB edge. The MOSI signal goes through a serial to parallel register forming internal 8-bit register SPI_received [7:0]. The control signals of waitsecondcycle, secondcyclewriteend, and secondcycleread are generated by a state machine which analyzes the SPI bus transaction and determines whether a SPI read or SPI write is to occur. Figure 9.8 shows the state machine which monitors and tracks the SPI bus protocol and generates the control logic for the SPI write and SPI read.

Referring to Fig. 9.8:

a. Reset will put the state machine into IDLE state
b. When CSB goes low, the state machine moves to firstcycle state. The state machine loops in this state, and the SPI_received[7:0] register in Fig. 9.7 receives the 8 bits of SPI data from MOSI. When the 8 bits of MOSI data transaction completes, CSB goes high.
c. When CSB goes high, the first 8 bits of SPI data completes its transaction and the state machine moves to waitsecondcycle state. The state machine stays in this state until it detects the second CSB goes low.

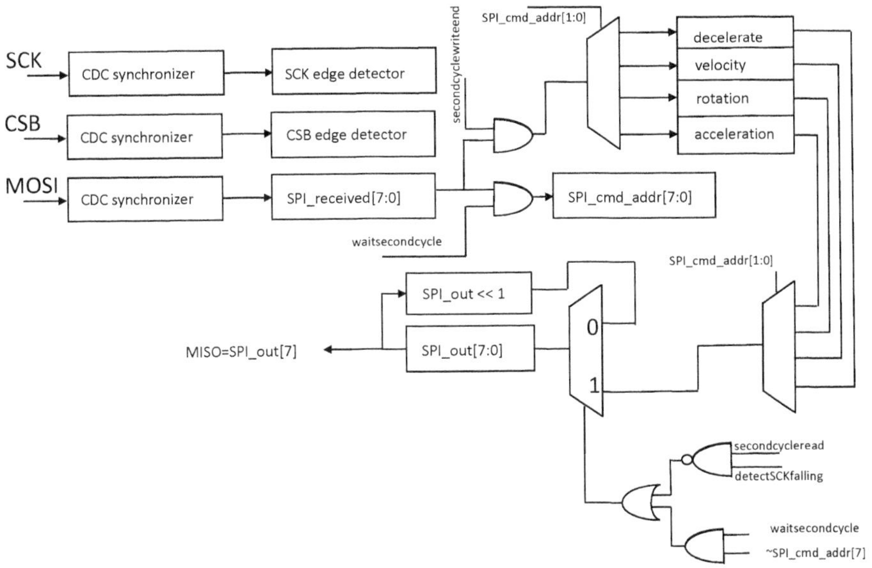

Fig. 9.7 Diagram showing block architecture Of SPI interface

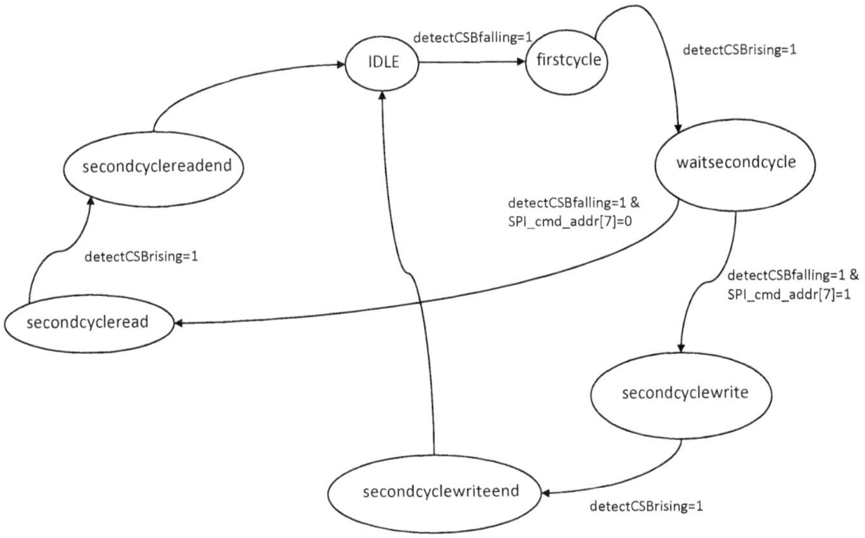

Fig. 9.8 Diagram showing state machine to analyze and track the SPI bus protocol

d. In waitsecondcycle state, the contents of SPI_received[7:0] is passed to SPI_cmd_addr[7:0]. The MSB of SPI_cmd_addr determines if the first byte of SPI command is a SPI read or SPI write. The lower bits [1:0] determines which of the four register the SPI command is to access.

e. In waitsecondcycle state, when CSB goes low, if SPI_cmd_addr[7] is 1, it is a SPI write command and the state machine moves to secondcyclewrite state. The state machine loops in this state, and the SPI_received[7:0] register in Fig. 9.7 receives the 8 bits of SPI data from MOSI. When the 8 bits of MOSI data transaction completes, CSB goes high, and the state machine moves to secondcyclewriteend state.
f. In the secondcyclewriteend state, the signal secondcyclewriteend goes high, and the data from SPI_received[7:0] is written into one of the four registers decelerate/velocity/rotation/acceleration based on SPI_cmd_addr[1:0].
g. On the next cycle after secondcyclewriteend, the state machine moves back to IDLE state.
h. In waitsecondcycle in (e), when CSB goes low, if SPI_cmd_addr[7] is 0, it is a SPI read command and the state machine moves to secondcycleread state. The state machine loops in this state, and the contents of one of the four registers decelerate/velocity/rotation/acceleration based on SPI_cmd_addr[1:0] are shifted out on every SCK through MISO.
i. When the 8 bits of MISO data transaction are complete, CSB goes high, and the state machine move to secondcyclereadend state.
j. On the next cycle after secondcyclereadend, the state machine moves back to IDLE state.

Based on the circuit diagram and state machine diagram shown in Figs. 9.7 and 9.8, Example 9.1 shows the Verilog code for the SPI slave interface and Example 9.2 shows the testbench for simulating and verifying the specified functionality.

Example 9.1 Example showing Verilog code: A SPI slave interface

```
`timescale 1ns/1ns
module spi_slave (clock, resetb, SCK, MISO, MOSI, CSB);

input clock, resetb;
input SCK, MOSI, CSB;
output MISO;

wire MISO;
reg SCKq1, SCKq2, SCKq3;
wire detectSCKrising, detectSCKfalling;
reg CSBq1, CSBq2, CSBq3;
wire detectCSBrising, detectCSBfalling;
wire CSBactive;
reg MOSIq1, MOSIq2;
reg [7:0] SPI_received;
reg [7:0] SPI_cmd_addr;
reg [7:0] SPI_out;
```

```
// internal spi_slave registers for read/write access
reg [7:0] decelerate, rotation, velocity, acceleration;

parameter [2:0] IDLE = 0,
              firstcycle = 1,
              waitsecondcycle = 2,
              secondcyclewrite = 3,
              secondcycleread = 4,
              secondcyclewriteend = 5,
              secondcyclereadend = 6;

reg [2:0] present_state, next_state;


always @ (posedge clock or negedge resetb)
begin
       if (~resetb)
              begin
                     MOSIq1 <= 0;
                     MOSIq2 <= 0;
              end
       else
              begin
                     MOSIq1 <= MOSI;
                     MOSIq2 <= MOSIq1;
              end
end

always @ (posedge clock or negedge resetb)
begin
       if (~resetb)
              begin
                     SCKq1 <= 0;
                     SCKq2 <= 0;
                     SCKq3 <= 0;
              end
       else
              begin
                     SCKq1 <= SCK;
                     SCKq2 <= SCKq1;
                     SCKq3 <= SCKq2;
              end
end

assign detectSCKrising = ~SCKq3 & SCKq2;
```

```
assign detectSCKfalling = SCKq3 & ~SCKq2;

always @ (posedge clock or negedge resetb)
begin
	if (~resetb)
		begin
			CSBq1 <= 0;
			CSBq2 <= 0;
			CSBq3 <= 0;
		end
	else
		begin
			CSBq1 <= CSB;
			CSBq2 <= CSBq1;
			CSBq3 <= CSBq2;
		end
end

assign detectCSBfalling = CSBq3 & ~CSBq2; // CSB falling means
start
assign detectCSBrising = ~CSBq3 & CSBq2; // CSB rising means end
assign CSBactive = ~CSBq2;

// latch in SPI data at MOSI
always @ (posedge clock or negedge resetb)
begin
	if (~resetb)
			SPI_received <= 0;
	else if (CSBactive & detectSCKrising)
			SPI_received <= {SPI_received[6:0], MOSIq2};
end

always @ (posedge clock or negedge resetb)
begin
	if (~resetb)
		present_state <= IDLE;
	else
		present_state <= next_state;
end

always @ (present_state or detectCSBfalling or detectCSBrising)
begin
	case (present_state)
		IDLE:
			begin
```

```
                    if (detectCSBfalling)
                        next_state = firstcycle;
                    else
                        next_state = IDLE;
                end
            firstcycle:
                begin
                    if (detectCSBrising)
                        next_state = waitsecondcycle;
                    else
                        next_state = firstcycle;
                end
            waitsecondcycle:
                begin
                    if (detectCSBfalling)
                        begin
                            if (SPI_cmd_addr[7]) //
SPI write
                                next_state =
                             secondcyclewrite;
                            else
                                next_state =
                             secondcycleread;
                        end
                    else
                        next_state = waitsecondcycle;
                end
            secondcyclewrite:
                begin
                    if (detectCSBrising)
                                    next_state =
secondcyclewriteend;
                    else
                        next_state = secondcyclewrite;
                end
            secondcycleread:
                begin
                    if (detectCSBrising)
                                    next_state =
secondcyclereadend;
                    else
                        next_state = secondcycleread;
                end
            secondcyclewriteend:
                next_state = IDLE;
```

```
            secondcyclereadend:
                    next_state = IDLE;
            default:
                    next_state = IDLE;
    endcase
end

always @ (posedge clock or negedge resetb)
begin
    if (~resetb)
            begin
                    SPI_cmd_addr <= 0;
            end
    else
            begin
                    if (present_state == waitsecondcycle)
                            SPI_cmd_addr <= SPI_received;
                    else if (present_state == IDLE)
                            SPI_cmd_addr <= 0;
            end
end

always @ (posedge clock or negedge resetb)
begin
    if (~resetb)
            begin
                    decelerate <= 0;
                    velocity <= 0;
                    rotation <= 0;
                    acceleration <= 0;
            end
    else
            begin
                    if (present_state == secondcyclewriteend)
                            begin
                                    case (SPI_cmd_addr[1:0])
                                            2'b00: decelerate <=
                                                    SPI_received;
                                                   2'b01: velocity <=
SPI_received;
                                                   2'b10: rotation <=
SPI_received;
                                            2'b11: acceleration <=
                                                    SPI_received;
                                    endcase
```

```
                        end
            end
end

always @ (posedge clock or negedge resetb)
begin
      if (~resetb)
            begin
                  SPI_out <= 0;
            end
      else
            begin
                  if (present_state == waitsecondcycle)
                        begin
                              if (~SPI_cmd_addr[7]) // SPI
read
                                    begin

case (SPI_cmd_addr[1:0])
                                                      2'b00:
SPI_out <=

decelerate;
                                                      2'b01:
SPI_out <=

velocity;
                                                      2'b10:
SPI_out <=
                                                    rotation;
                                                      2'b11:
SPI_out <=

acceleration;
                                          endcase
                                    end
                        end
                  else if (detectSCKfalling & (present_state ==
                           secondcycleread))
                        SPI_out <= {SPI_out[6:0], 1'b0};
            end
end
```

```
assign MISO = SPI_out[7];  // SPI sends out MSB first


endmodule
```

Example 9.2 Example showing Verilog code for a testbench simulating SPI slave interface

```
`timescale 1ns/1ns
module spi_slave_tb ();

reg CSB;
reg SCK;
reg SDI;
wire SDO;
reg clock;
reg resetb;
integer SPI_delay = 100;
integer system_delay = 100;
integer CSB_stabilize_delay = 1000;
integer j;
reg [7:0] SPIwritedata;
reg [7:0] SPIreadaddr;
reg [7:0] read_data;

initial
begin
       clock = 0;
       forever #25 clock = ~clock; // 20 MHz clock
end

initial
begin
       resetb = 1;
       #system_delay;
       resetb = 0;
       #system_delay;
       resetb = 1;
       #system_delay;

      SPIwritedata = 8'h80;    // command for write to address 00h
                                    // with data aah
       SPI_8bit_write;
       #CSB_stabilize_delay;
```

```
    SPIwritedata = 8'haa;
    SPI_8bit_write;
    #CSB_stabilize_delay;

    SPIwritedata = 8'h81;    // command for write to address 01h
                                 // with data abh
    SPI_8bit_write;
    #CSB_stabilize_delay;
    SPIwritedata = 8'hab;
    SPI_8bit_write;
    #CSB_stabilize_delay;

    SPIwritedata = 8'h82;    // command for write to address 02h
                                 // with data ach
    SPI_8bit_write;
    #CSB_stabilize_delay;
    SPIwritedata = 8'hac;
    SPI_8bit_write;
    #CSB_stabilize_delay;

    SPIwritedata = 8'h83;    // command for write to address 03h
                                 // with data adh
    SPI_8bit_write;
    #CSB_stabilize_delay;
    SPIwritedata = 8'had;
    SPI_8bit_write;
    #CSB_stabilize_delay;

        #CSB_stabilize_delay;       //  delay  and  wait  CSB  to
stabilize


    // do read from address 00h
    SPIreadaddr = 8'h00;     // command for read address 00h
    SPI_8bit_read;
    #CSB_stabilize_delay;    // wait to stabilize CSB

    // do read from address 01h
    SPIreadaddr = 8'h01;     // command for read address 01h
    SPI_8bit_read;
    #CSB_stabilize_delay;    // wait to stabilize CSB

    // do read from address 02h
    SPIreadaddr = 8'h02;     // command for read address 02h
    SPI_8bit_read;
```

```
        #CSB_stabilize_delay;     // wait to stabilize CSB

        // do read from address 03h
        SPIreadaddr = 8'h03;      // command for read address 03h
        SPI_8bit_read;
        #CSB_stabilize_delay;     // wait to stabilize CSB

        // bring back CSB = 1
        CSB = 1;
        #CSB_stabilize_delay;

        #1000;
        $stop;
end

task SPI_8bit_write;
begin
        CSB = 1;
        SCK = 0;
        SDI = 0;
        #SPI_delay;
        CSB = 0;
        for (j=8; j>0; j=j-1)
        begin
                SCK = 0;
                SDI = SPIwritedata[j-1];
                #SPI_delay;
                SCK = 1;
```

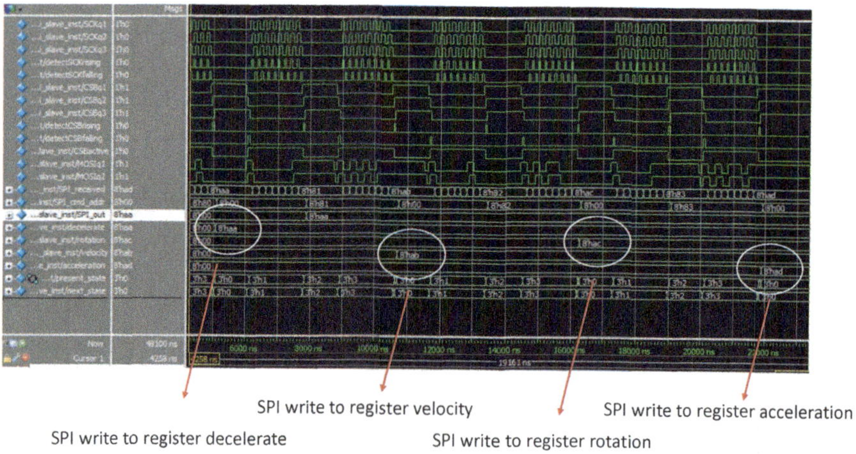

Fig. 9.9 Diagram showing simulation of SPI write

```
                #SPI_delay;
                SCK = 0;
        end
        #SPI_delay;
        CSB = 1;
end
endtask

task SPI_8bit_read;
```

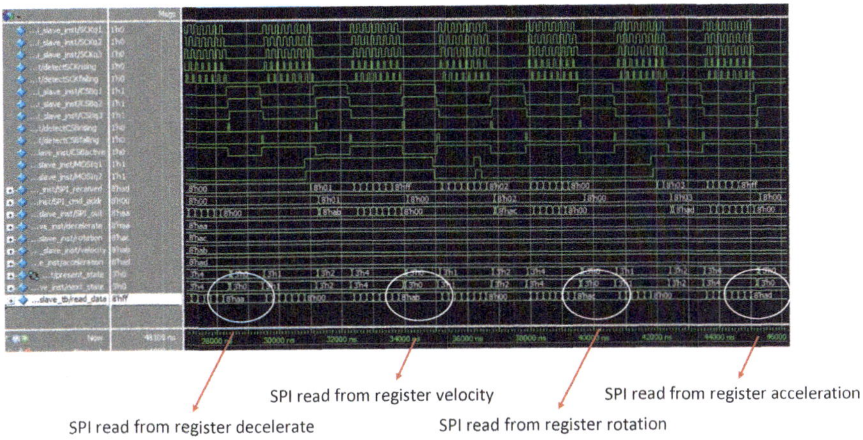

Fig. 9.10 Diagram showing simulation of SPI read

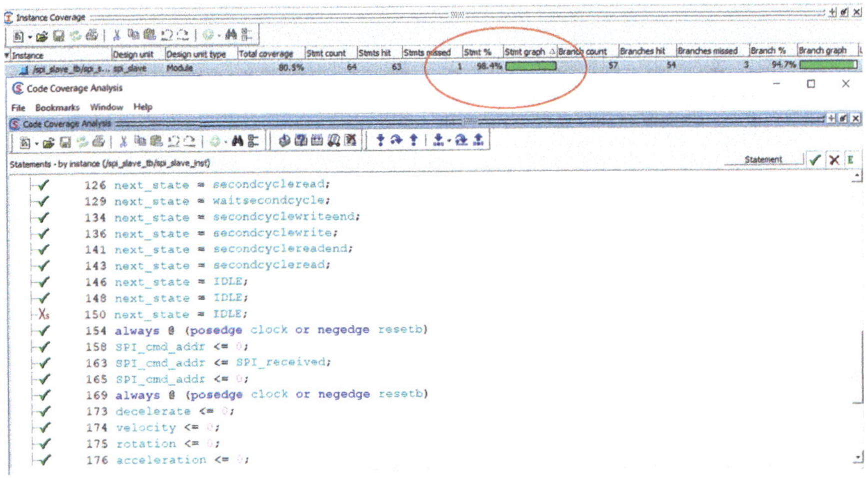

Fig. 9.11 Diagram showing code coverage for statement

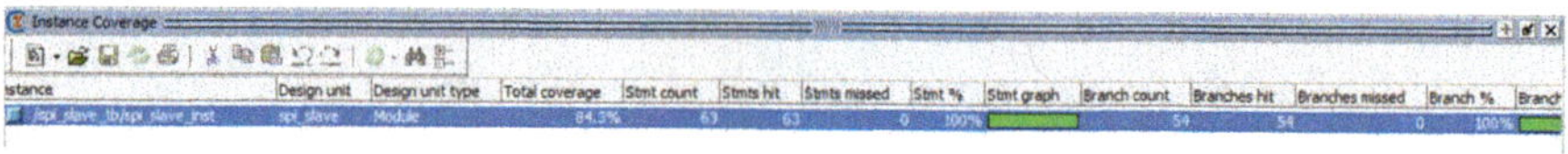

Fig. 9.12 Diagram showing code coverage for statement achieving 100%

```
begin
        CSB = 1;
        SCK = 0;
        SDI = 0;
        #SPI_delay;
        CSB = 0;
        for (j=8; j>0; j=j-1)
        begin
                SCK = 0;
                SDI = SPIreadaddr[j-1];
                #SPI_delay;
                SCK = 1;
                #SPI_delay;
                SCK = 0;
        end
        #SPI_delay;
        CSB = 1;
       #CSB_stabilize_delay;

        // read data
        CSB = 1;
        SCK = 0;
        #SPI_delay;
        CSB = 0;
        for (j=8; j>0; j=j-1)
        begin
                SCK = 0;
                #SPI_delay;
                SCK = 1;
                #SPI_delay;
                SCK = 0;
        end
        #SPI_delay;
        CSB = 1;
end
endtask

always @ (negedge SCK)
begin
```

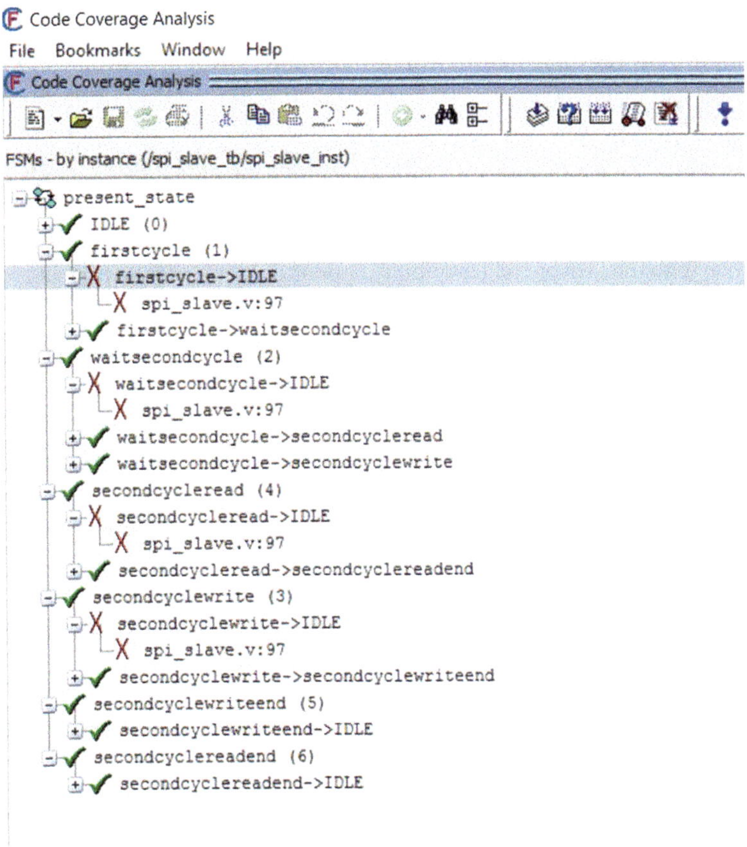

Fig. 9.13 Diagram showing missing transitions in FSM code coverage

```
        read_data <= {read_data[6:0], SDO};
end

spi_slave    spi_slave_inst    (.clock(clock),    .resetb(resetb),
.SCK(SCK), .MISO(SDO), .MOSI(SDI), .CSB(CSB));


endmodule
```

Simulation of testbench of Example 9.2 is shown in Fig. 9.9 for SPI write to registers and Fig. 9.10 for SPI read from registers.

Figures 9.9 and 9.10 show the simulation of SPI read and SPI write with the correct functionality of writing data from SPI bus to the internal registers and functionality of read data from internal registers to SPI bus. However the RTL Verilog code of the SPI slave interface shown in Example 9.1 may not be fully exercised by the testbench shown in Example 9.2; thus the usage of code coverage to identify which parts of RTL code is not yet exercised.

Figure 9.11 shows the statement code coverage of 98.5% for the simulation using Mentor Graphics' ModelSim simulator.

Referring to Fig. 9.11, the missing statement in the coverage is in line 150 of the RTL Verilog code of Example 9.1. This missing statement in the code coverage can be ignored due to its usage in the RTL code for the default of the case statement with all possible conditions of the case statement being specified. By recompiling the RTL code with the "coverexcludedefault" option for vlog in ModelSim and resimulating the testbench achieves a 100% statement coverage as shown in Fig. 9.12.

- vlog -coverexcludedefault +cover=sbceft spi_slave.v
- vlog -coverexcludedefault +cover=sbceft spi_slave_tb.v
- vsim -coverage spi_slave_tb

The coverage report shown in Example 9.3 shows that the FSM (finite-state machine) transitions have a 66.6% coverage; FEC (focused expression coverage) conditions have a 60% coverage and toggle coverage at 62.6%. The statement coverage, branch coverage, FSM states coverage, and FEC expression coverage are at 100%. From this coverage report, the testbench shown in Example 9.2 must be enhanced to increase the coverage of FSM transitions, FEC conditions, and toggle nodes.

Example 9.3 Example showing coverage report with inadequate coverage on FSM, FEC, and toggle

```
Coverage Report Summary Data by File

====================================================================
====
=== File: spi_slave.v
====================================================================
====
    Enabled Coverage              Active      Hits    Misses % Covered
    ----------------              ------      ----    ------ ---------
    Stmts                             63        63         0     100.0
    Branches                          54        54         0     100.0
    FEC Condition Terms                5         3         2      60.0
    FEC Expression Terms               8         8         0     100.0
    FSMs                                                          83.3
        States                         7         7         0     100.0
        Transitions                   12         8         4      66.6
    Toggle Bins                      150        94        56      62.6
```

Enhancing Testbench to Increase Code Coverage

The FSM states coverage is at 100% as the testbench cycles the state machine through all the states. However the FSM transitions coverage is only at 66.6% indicating that some of the transitions in the state machine are not covered. Nonsimulated state transitions may lead to potential bugs in the design; thus additional testbench conditions must be added to increase the state machine transition coverage.

Figure 9.13 shows the missing transition coverage as firstcycle → IDLE, waitsecondcycle → IDLE, secondcycleread → IDLE, and secondcyclewrite → IDLE.

Referring to the RTL code of Example 9.1, there is no RTL Verilog path from firstcycle → IDLE as the RTL only have if-else statement to remain in firstcycle state or to move to waitsecondcycle state. However the code coverage is showing a missing transition check for those states. This is due to the FSM coverage checking for all possible state changes in the FSM, and this includes reset when the state machine is functioning. A reset may occur when the state machine is in firstcycle thus firstcycle → IDLE. Similarly a reset may occur when the state machine is in waitsecondcycle thus waitsecondcycle → IDLE.

The testbench Verilog code shown in Example 9.4 is the additional code added to the testbench of Example 9.2 to verify the missing coverage of FSM transitions from firstcycle → IDLE, waitsecondcycle → IDLE, secondcyclewrite → IDLE, and secondcycleread → IDLE. What this additional testbench code does is to drive the reset when the FSM is in those states.

Example 9.4 Example showing additional Verilog testbench code to simulate missing transitions in FSM code coverage

```
// coverage from firstcycle -> IDLE
SPIwritedata = 8'h80;    // command for write to address 00h
                           // with data aah
SPI_8bit_write_reset;

// coverage from waitsecondcycle -> IDLE
SPIwritedata = 8'h80;    // command for write to address 00h
                           // with data aah
SPI_8bit_write;
#CSB_stabilize_delay;
resetb = 1;
#system_delay;
resetb = 0;
#system_delay;
resetb = 1;
#system_delay;

// coverage from secondcyclewrite -> IDLE
```

```
        SPIwritedata = 8'h80;    // command for write to address 00h
                                    // with data aah
        SPI_8bit_write;
        #CSB_stabilize_delay;
        SPIwritedata = 8'hab;
        SPI_8bit_write_reset;

        // coverage from secondcycleread -> IDLE
        // do read from address 00h
        SPIreadaddr = 8'h00; // command for read address 00h
        SPI_8bit_read_reset;

task SPI_8bit_write_reset;
begin
        CSB = 1;
        SCK = 0;
        SDI = 0;
        #SPI_delay;
        CSB = 0;
        for (j=8; j>0; j=j-1)
        begin
                SCK = 0;
                SDI = SPIwritedata[j-1];
                #SPI_delay;
                SCK = 1;
                #SPI_delay;
                SCK = 0;
        end
        #SPI_delay;
        resetb = 1;
        #system_delay;
        resetb = 0;
        CSB = 1;
        #system_delay;
        resetb = 1;
        #system_delay;
end
endtask

task SPI_8bit_read_reset;
begin
        CSB = 1;
        SCK = 0;
        SDI = 0;
        #SPI_delay;
```

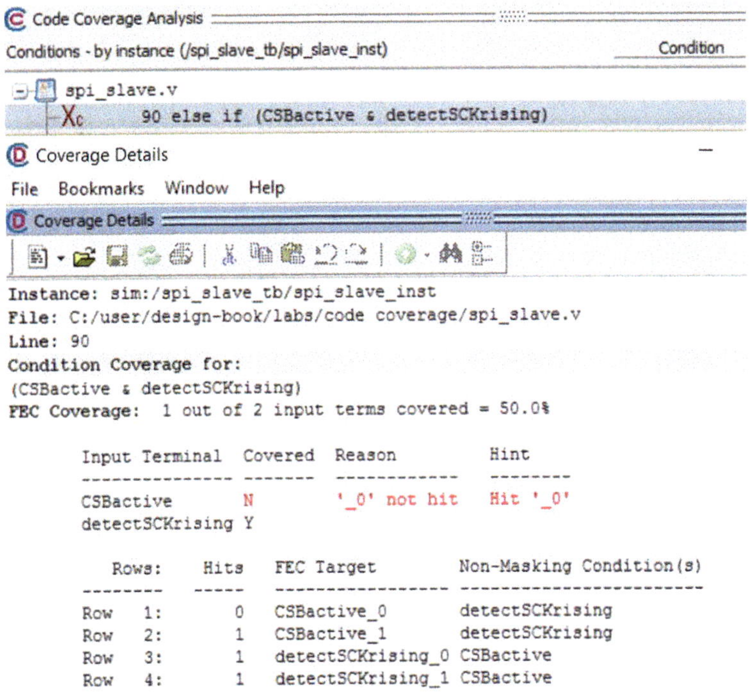

Fig. 9.14 Diagram showing missing condition coverage

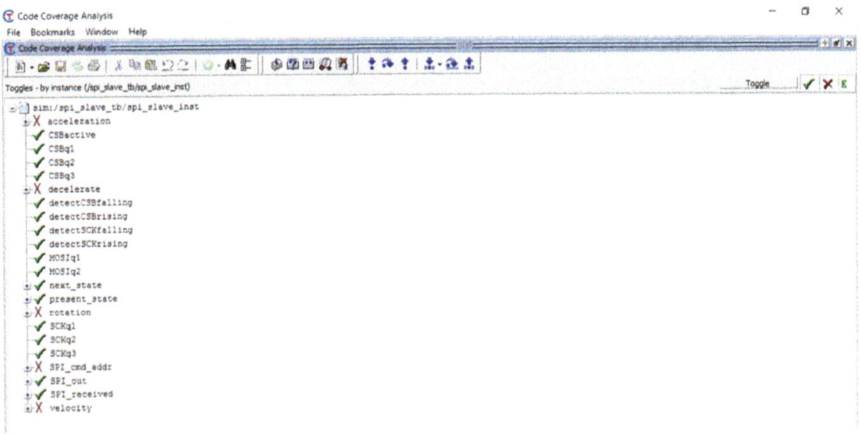

Fig. 9.15 Diagram showing missing toggle coverage

```
        CSB = 0;
        for (j=8; j>0; j=j-1)
        begin
                SCK = 0;
                SDI = SPIreadaddr[j-1];
                #SPI_delay;
                SCK = 1;
                #SPI_delay;
                SCK = 0;
        end
        #SPI_delay;
        CSB = 1;
       #CSB_stabilize_delay;

        // read data
        CSB = 1;
        SCK = 0;
        #SPI_delay;
        CSB = 0;
        for (j=8; j>0; j=j-1)
        begin
                SCK = 0;
                #SPI_delay;
                SCK = 1;
                #SPI_delay;
                SCK = 0;
        end
        #SPI_delay;
        #system_delay;
        resetb = 0;
        CSB = 1;
        #system_delay;
        resetb = 1;
        #system_delay;
end
endtask
```

Figure 9.14 shows the missing condition coverage which the testbench in Example 9.2 did not cover. It shows the condition (CSBactive & detectSCKrising) did not cover detectSCKrising when CSBactive is low. As the SPI master only drives SCK during CSB low, this uncovered condition can be waived. However, for completeness of code coverage, an additional testbench stimulus can be added as shown in Example 9.5.

The additional Verilog testbench shown in Example 9.5 drives the SCK signal while CSB is high, thus covering the condition of detectSCKrising when CSBactive is low.

Example 9.5 Example showing additional Verilog testbench code to simulate missing condition coverage

```
// coverage (CSBactive & detectSCKrising)
resetb = 1;
CSB = 1;
SCK = 0;
#system_delay;
CSB = 0;
#SPI_delay;
SCK = 1;
#SPI_delay;
SCK = 0;
#SPI_delay;
SCK = 1;
CSB = 1;
#SPI_delay;
SCK = 0;
#SPI_delay;
CSB = 0;
#SPI_delay;
SCK = 1;
#SPI_delay;
SCK = 0;
#SPI_delay;
SCK = 1;
#SPI_delay;
SCK = 0;
CSB = 1;
```

The diagram shown in Fig. 9.15 is the missing toggle coverage which is not covered in the testbench of Example 9.2. The registers acceleration, deceleration, rotation, velocity, and SPI_cmd_addr are not fully exercised to have 100% toggle coverage.

To increase the toggle coverage, additional testbench stimulus can be added to write to the registers as well as the SPI_cmd_addr as shown in Example 9.6.

Example 9.6 Example showing additional Verilog testbench code to increase toggle coverage

```
// increase toggle coverage for the 4 registers
```

```
        for (k=128; k<=131; k=k+1)
                begin
                        SPIwritedata = k; // command for write to
                                                  // address 00h with data
00h
                        SPI_8bit_write;
                        #CSB_stabilize_delay;
                        SPIwritedata = 8'h00;
                        SPI_8bit_write;
                        #CSB_stabilize_delay;
                          SPIwritedata = k; // command for write to
address
                                                   // 00h with data ffh
                        SPI_8bit_write;
                        #CSB_stabilize_delay;
                        SPIwritedata = 8'hff;
                        SPI_8bit_write;
                        #CSB_stabilize_delay;
                          SPIwritedata = k; // command for write to
address
                                                   // 00h with data 00h
                        SPI_8bit_write;
                        #CSB_stabilize_delay;
                        SPIwritedata = 8'h00;
                        SPI_8bit_write;
                        #CSB_stabilize_delay;
                end

        // increase toggle coverage for SPI_cmd_addr[7:2]
        for (k=0; k<=255; k=k+1)
                begin

SPIwritedata = k; // cycle through SPI_cmd_addr
                        SPI_8bit_write;
                        #CSB_stabilize_delay;
                        SPIwritedata = 8'h00;
                        SPI_8bit_write;
                        #CSB_stabilize_delay;
                end
        #CSB_stabilize_delay; // delay and wait CSB to stabilize
```

With the additional Verilog testbench code of Examples 9.4, 9.5, and 9.6, the final testbench shown in Example 9.7 can achieve 100% code coverage as shown in Example 9.8 of the code coverage report.

Example 9.7 Example showing Verilog code for a testbench simulating SPI slave interface

```
`timescale 1ns/1ns
module spi_slave_tb ();

reg CSB;
reg SCK;
reg SDI;
wire SDO;
reg clock;
reg resetb;
integer SPI_delay = 100;
integer system_delay = 100;
integer CSB_stabilize_delay = 1000;
integer j,k;
reg [7:0] SPIwritedata;
reg [7:0] SPIreadaddr;
reg [7:0] read_data;

initial
begin
      clock = 0;
      forever #25 clock = ~clock; // 20 MHz clock
end

initial
begin
      resetb = 1;
      #system_delay;
      resetb = 0;
      #system_delay;
      resetb = 1;
      #system_delay;

      SPIwritedata = 8'h80; // command for write to address 00h
with
                                      // data aah
      SPI_8bit_write;
      #CSB_stabilize_delay;
      SPIwritedata = 8'haa;
      SPI_8bit_write;
      #CSB_stabilize_delay;

      SPIwritedata = 8'h81; // command for write to address 01h
with
```

```
                                  // data abh
     SPI_8bit_write;
     #CSB_stabilize_delay;
     SPIwritedata = 8'hab;
     SPI_8bit_write;
     #CSB_stabilize_delay;

     SPIwritedata = 8'h82; // command for write to address 02h
with
                              // data ach
     SPI_8bit_write;
     #CSB_stabilize_delay;
     SPIwritedata = 8'hac;
     SPI_8bit_write;
     #CSB_stabilize_delay;

     SPIwritedata = 8'h83; // command for write to address 03h
with
                                  // data adh
     SPI_8bit_write;
     #CSB_stabilize_delay;
     SPIwritedata = 8'had;
     SPI_8bit_write;
     #CSB_stabilize_delay;

     #CSB_stabilize_delay; // delay and wait CSB to stabilize


     // do read from address 00h
     SPIreadaddr = 8'h00; // command for read address 00h
     SPI_8bit_read;
     #CSB_stabilize_delay; // wait to stabilize CSB

     // do read from address 01h
     SPIreadaddr = 8'h01; // command for read address 01h
     SPI_8bit_read;
     #CSB_stabilize_delay; // wait to stabilize CSB

     // do read from address 02h
     SPIreadaddr = 8'h02; // command for read address 02h
     SPI_8bit_read;
     #CSB_stabilize_delay; // wait to stabilize CSB

     // do read from address 03h
     SPIreadaddr = 8'h03; // command for read address 03h
```

```
    SPI_8bit_read;
    #CSB_stabilize_delay; // wait to stabilize CSB

    // bring back CSB = 1
    CSB = 1;
    #CSB_stabilize_delay;

    resetb = 1;
    #system_delay;
    resetb = 0;
    #system_delay;
    resetb = 1;
    #system_delay;

    // coverage from firstcycle -> IDLE
    SPIwritedata = 8'h80; // command for write to address 00h
with
                            // data aah
    SPI_8bit_write_reset;

    // coverage from waitsecondcycle -> IDLE
    SPIwritedata = 8'h80; // command for write to address 00h
with
                            // data aah
    SPI_8bit_write;
    #CSB_stabilize_delay;
    resetb = 1;
    #system_delay;
    resetb = 0;
    #system_delay;
    resetb = 1;
    #system_delay;

    // coverage from secondcyclewrite -> IDLE
    SPIwritedata = 8'h80; // command for write to address 00h
with
                                // data aah
    SPI_8bit_write;
    #CSB_stabilize_delay;
    SPIwritedata = 8'hab;
    SPI_8bit_write_reset;

    // coverage from secondcycleread -> IDLE
    // do read from address 00h
    SPIreadaddr = 8'h00; // command for read address 00h
```

```
SPI_8bit_read_reset;

// coverage (CSBactive & detectSCKrising)
resetb = 1;
CSB = 1;
SCK = 0;
#system_delay;
CSB = 0;
#SPI_delay;
SCK = 1;
#SPI_delay;
SCK = 0;
#SPI_delay;
SCK = 1;
CSB = 1;
#SPI_delay;
SCK = 0;
#SPI_delay;
CSB = 0;
#SPI_delay;
SCK = 1;
#SPI_delay;
SCK = 0;
#SPI_delay;
SCK = 1;
#SPI_delay;
SCK = 0;
CSB = 1;

#system_delay;
resetb = 1;
#system_delay;
resetb = 0;
#system_delay;
resetb = 1;
#system_delay;

// increase toggle coverage for the 4 registers


for (k=128; k<=131; k=k+1)
        begin
                SPIwritedata = k; // command for write to
                                            // address 00h with data
00h
```

```
                    SPI_8bit_write;
                    #CSB_stabilize_delay;
                    SPIwritedata = 8'h00;
                    SPI_8bit_write;
                    #CSB_stabilize_delay;
                      SPIwritedata = k; // command for write to
address
                                          // 00h with data ffh
                    SPI_8bit_write;
                    #CSB_stabilize_delay;
                    SPIwritedata = 8'hff;
                    SPI_8bit_write;
                    #CSB_stabilize_delay;
                      SPIwritedata = k; // command for write to
address
                                          // 00h with data 00h
                    SPI_8bit_write;
                    #CSB_stabilize_delay;
                    SPIwritedata = 8'h00;
                    SPI_8bit_write;
                    #CSB_stabilize_delay;
               end

        // increase toggle coverage for SPI_cmd_addr[7:2]
        for (k=0; k<=255; k=k+1)
               begin

SPIwritedata = k; // cycle through SPI_cmd_addr
                    SPI_8bit_write;
                    #CSB_stabilize_delay;
                    SPIwritedata = 8'h00;
                    SPI_8bit_write;
                    #CSB_stabilize_delay;
               end
        #CSB_stabilize_delay; // delay and wait CSB to stabilize

        #1000;


        $stop;
end

task SPI_8bit_write;
begin
        CSB = 1;
```

```
        SCK = 0;
        SDI = 0;
        #SPI_delay;
        CSB = 0;
        for (j=8; j>0; j=j-1)
        begin
                SCK = 0;
                SDI = SPIwritedata[j-1];
                #SPI_delay;
                SCK = 1;
                #SPI_delay;
                SCK = 0;
        end
        #SPI_delay;
        CSB = 1;
end
endtask

task SPI_8bit_write_reset;
begin
        CSB = 1;
        SCK = 0;
        SDI = 0;
        #SPI_delay;
        CSB = 0;
        for (j=8; j>0; j=j-1)
        begin
                SCK = 0;
                SDI = SPIwritedata[j-1];
                #SPI_delay;
                SCK = 1;
                #SPI_delay;
                SCK = 0;
        end
        #SPI_delay;
        resetb = 1;
        #system_delay;
        resetb = 0;
        CSB = 1;
        #system_delay;
        resetb = 1;
        #system_delay;
end
endtask
```

```
task SPI_8bit_read;
begin
        CSB = 1;
        SCK = 0;
        SDI = 0;
        #SPI_delay;
        CSB = 0;
        for (j=8; j>0; j=j-1)
        begin
                SCK = 0;
                SDI = SPIreadaddr[j-1];
                #SPI_delay;
                SCK = 1;
                #SPI_delay;
                SCK = 0;
        end
        #SPI_delay;
        CSB = 1;
       #CSB_stabilize_delay;

        // read data
        CSB = 1;
        SCK = 0;
        #SPI_delay;
        CSB = 0;
        for (j=8; j>0; j=j-1)
        begin
                SCK = 0;
                #SPI_delay;
                SCK = 1;
                #SPI_delay;
                SCK = 0;
        end
        #SPI_delay;
        CSB = 1;
end
endtask

task SPI_8bit_read_reset;
begin
        CSB = 1;
        SCK = 0;
        SDI = 0;
        #SPI_delay;
        CSB = 0;
```

```
        for (j=8; j>0; j=j-1)
        begin
                SCK = 0;
                SDI = SPIreadaddr[j-1];
                #SPI_delay;
                SCK = 1;
                #SPI_delay;
                SCK = 0;
        end
        #SPI_delay;
        CSB = 1;
       #CSB_stabilize_delay;

        // read data
        CSB = 1;
        SCK = 0;
        #SPI_delay;
        CSB = 0;
        for (j=8; j>0; j=j-1)
        begin
                SCK = 0;
                #SPI_delay;
                SCK = 1;
                #SPI_delay;
                SCK = 0;
        end
        #SPI_delay;
        #system_delay;
        resetb = 0;
        CSB = 1;
        #system_delay;
        resetb = 1;
        #system_delay;
end
endtask

always @ (negedge SCK)
begin
        read_data <= {read_data[6:0], SDO};
end

spi_slave   spi_slave_inst   (.clock(clock),   .resetb(resetb),
.SCK(SCK), .MISO(SDO), .MOSI(SDI), .CSB(CSB));


endmodule
```

Example 9.8 Example showing coverage report with 100% coverage

```
Coverage Report Summary Data by File

===================================================================
====
=== File: spi_slave.v
===================================================================
====
    Enabled Coverage            Active       Hits    Misses % Covered
    ----------------            ------       ----    ------ ---------
    Stmts                           63         63         0     100.0
    Branches                        54         54         0     100.0
    FEC Condition Terms              4          4         0     100.0
    FEC Expression Terms             8          8         0     100.0
    FSMs                                                        100.0
        States                       7          7         0     100.0
        Transitions                 12         12         0     100.0
    Toggle Bins                    150        150         0     100.0
```

With 100% code coverage, the Verilog testbench fully exercise all the RTL code of SPI slave. However it is important to note that RTL code coverage is not functional coverage. Code coverage ensures that there is adequate coverage of all the RTL code being exercised by the testbench. Any part of RTL code that is not exercised by a testbench is a potential source of RTL bug. It is a good RTL coding practice to ensure 100% code coverage if possible.

References

1. Estimating MTBF of Multi Stage Synchronizers, D. Zar, T. Chaney, J. Cox, S. Beer, R. Ginosar, Blendics Inc., St. Louis, Missouri, EE Dept., Technion Israel Institute of Technology, Haifa, Israel
2. Clock Domain Crossing (CDC) Design & Verification Techniques Using SystemVerilog, Clifford E. Cummings, Sunburst Design, Inc., SNUG 2008
3. Five Steps to Quality CDC Verification, Ping Yeung, Mentor Graphics, Advanced Verification White Paper
4. Design and Measurement of Synchronizers, Jun Zhou, School of Electrical, Electronic & Computer Engineering, Newcastle University, Technical Report Series, NCL-EECE-MSD-TR-2008-138, November 2008
5. Synchronization in Digital Logic Circuits, Ryan Donohue
6. Designs with multiple clock domains: New tools avoid clock skew and reduce pattern counts, Geir Eide, Mentor Graphics Corp., EETimes, January 23 2002
7. FPGA For Fun, www.fpga4fun.com
8. Logic Built-in Self Test (LBIST) is Back – But Not for Manufacturing Test, 2012, Cadence Corp.
9. LBISTArchitect™ Process Guide, December 2015, © Mentor Graphics
10. MBISTArchitect™ Process Guide, February 2012.1, © Mentor Graphics
11. Introduction to Low Level Programming and Computer Organization, Kitty Reeves, Computer Science and Engineering, Ohio State University, 2012
12. Code Coverage Fundamentals, October 7, 2013, Sini Balakrishnan, VLSI Pro, http://vlsi.pro/code-coverage-fundamentals/
13. Mentor Graphics' Modelsim SE User Manual, Code Coverage
14. VHDL Coding and Logic Synthesis with Synopsys, Weng Fook Lee, Academic Press, ISBN-13: 978-0123886712, ISBN-10: 0123886716
15. Verilog Coding for Logic Synthesis, Weng Fook Lee, Wiley Interscience, ISBN-13: 978-0471429760, ISBN-10: 0471429767
16. VLIW Microprocessor Hardware Design: On ASIC and FPGA, Weng Fook Lee, McGraw-Hill Education, ISBN-10: 0071497021, ISBN-13: 978-0071497022

W. F. Lee, *Learning from VLSI Design Experience*,
https://doi.org/10.1007/978-3-030-03238-8

Index

W. F. Lee, *Learning from VLSI Design Experience*,
https://doi.org/10.1007/978-3-030-03238-8

MIX
Papier aus verantwortungsvollen Quellen
Paper from responsible sources
FSC® C105338

Printed by Books on Demand, Germany